W9-CUA-667

AutoCAD® 2013
3D Modeling

LICENSE, DISCLAIMER OF LIABILITY, AND LIMITED WARRANTY

AUTOCAD® 2013
3D MODELING

By
Munir M. Hamad
Autodesk® Approved Instructor

MERCURY LEARNING AND INFORMATION
Dulles, Virginia
Boston, Massachusetts
New Delhi

Publisher: David Pallai

MERCURY LEARNING AND INFORMATION LLC
22841 Quicksilver Drive
Dulles, VA 20166
info@merclearning.com
www.merclearning.com
1-800-758-3756

This book is printed on acid-free paper.

Munir M. Hamad, AUTOCAD® 2013 3D MODELING.
ISBN: 9781936420414

The publisher recognizes and respects all marks used by companies, manufacturers, and developers as a means to distinguish their products. All brand names and product names mentioned in this book are trademarks or service marks of their respective companies. Any omission or misuse (of any kind) of service marks or trademarks, etc. is not an attempt to infringe on the property of others.

Library of Congress Control Number: 2012936942

12131432

Our titles are available for adoption, license, or bulk purchase by institutions, corporations, etc. For additional information, please contact the Customer Service Dept. at 1-800-758-3756 (toll free).

The sole obligation of MERCURY LEARNING AND INFORMATION to the purchaser is to replace the disc, based on defective materials or faulty workmanship, but not based on the operation or functionality of the product.

CONTENTS

PREFACE

- Since its inception, AutoCAD has enjoyed a very wide user base and has been the most widely used CAD software since the 1980s. This popularity is due to its logic and simplicity, which makes it very easy to learn.
- AutoCAD has evolved throughout the years and has become comprehensive software, addressing all aspects of engineering drafting and designing.
- This book addresses the 3D environment of AutoCAD and teaches the reader how to create and edit 3D objects in AutoCAD and produce real-world images out of them. This book is ideal for advanced AutoCAD users and college students that have already studied the basics of AutoCAD 2013.
- This book is not a replacement for the manual(s) that comes with the software but is considered complementary with its practices and projects, meant to strengthen the knowledge gained and solidify the techniques discussed.
- Solving all practices and projects is essential because AutoCAD is practical and not theoretical.
- At the end of each chapter the reader will find "Chapter Review Questions." These are the same sort of questions you might see in an Autodesk exam. The answers to the odd questions are included at the end of each chapter to check your work.
- This book contains four different engineering projects. Solving these projects will allow the reader to master the knowledge needed to land a job in today's market. These projects are presented in metric and imperial units.

INTRODUCTION

This book discusses the 3D environment and commands in AutoCAD 2013. Many AutoCAD users don't realize the 3D capabilities of AutoCAD, so they often end up using other software (such as 3ds Max or Maya) to build 3D models, create lights and materials, and create still rendered images and animation. This book clears up this misunderstanding and introduces the real power of AutoCAD in 3D.

At the completion of this book, the reader will be able to:

- Understand the AutoCAD 3D environment
- Create 3D objects using solids
- Create 3D objects using meshes
- Create 3D objects using surfaces
- Create complex solids and surfaces
- Edit solids and use other general 3D modifying commands
- Convert 3D objects from one type to another
- Create 2D/3D sections from 3D solids
- Use Model Documentation
- Create 3D DWF files
- Create cameras and lights
- Assign materials and create rendered images
- Use and create visual styles
- Create animation files

ABOUT THE DVD

The DVD included with this book contains:

- A link to the AutoCAD 2013 trial version (http://usa.autodesk.com/autocad/trial/), which will last for 30 days starting from the day of installation. This version will help you solve all the exercises and workshops in this book.
- The practices files, which will be your starting point to solve all exercises and workshops in the book.
- Copy the folder named "Practices and Projects" onto the hard drive of your computer. In the Project folder, you will find two folders; the first is called "Metric" for metric units projects and the second one is called "Imperial" for imperial units projects.

PREREQUISITES

- The author assumes the reader has experience using computers and the Windows operating system. The reader should know basic commands and be able to create new files, open existing ones, save and use save as well as close and exit the software. These fundamentals are not covered in this book.
- AutoCAD 2013 comes with a dark gray background, but the screenshots in this book have been changed to white to create clearer illustrations.

1 AUTOCAD 3D BASICS

In This Chapter

◇ How to use the 3D interface
◇ How to view a 3D model
◇ How to use the UCS commands

1.1 RECOGNIZING THE 3D ENVIRONMENT

- You will use the same software to draw 2D objects and 3D models, but AutoCAD separates the two environments by applying two rules:
  - You will use 3D template files, ***acad3D.dwt*** and ***acadiso3D.dwt***, to create the 3D environment and start building the desired 3D model by viewing things specifically for 3D in AutoCAD.
  - You will use either the ***3D Basic*** workspace or the ***3D Modeling*** workspace. In this book we will always use the 3D Modeling workspace to reach our commands because it includes tabs and panels that contain the AutoCAD 3D commands.

- The following illustrates these two concepts:

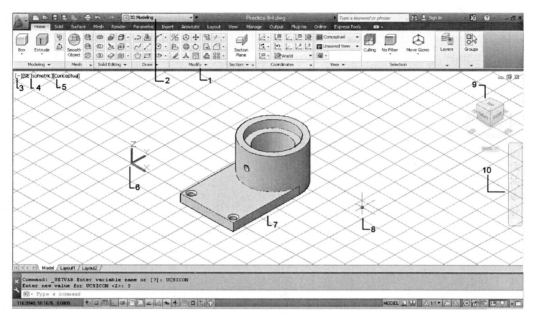

1. 3D Tabs and Panels	5. Visual Style Controls	8. 3D Crosshairs
2. Workspace	6. UCS Icon	9. ViewCube
3. Viewport Controls	7. 3D Model	10. Navigation Bar
4. View Controls		

1.2 VIEWING YOUR MODEL IN THE 3D ENVIRONMENT

1.2.1 Using Preset Views

- Viewing a 3D model in AutoCAD is simple and effective. The first method of viewing your model is to use the ten preset views built into AutoCAD. Assuming:
  - Positive Y-axis = North and Front
  - Negative Y-axis = South and Back
  - Positive X-axis = East and Right

- Negative X-axis = West and Left
- Positive Z-axis = Top
- Negative Z-axis = Bottom
- AutoCAD provides six 2D orthographic views and four 3D views. The six 2D orthographic views are:
  - Top
  - Bottom
  - Left
  - Right
  - Front
  - Back
- And the four 3D views are (look up to know each direction):
  - SW (South West) Isometric
  - SE (South East) Isometric
  - NE (North East) Isometric
  - NW (North West) Isometric
- To use these preset views, go to the top-left part of the screen, select **View Controls**, and then select the desired preset view. See the following illustration:

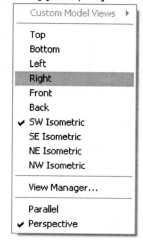

- You can find these commands using the **Home** tab and the **View** panel, but the above method is easier.

1.2.2 Using the ViewCube

- At the upper-right corner of each drawing, AutoCAD will show the ViewCube. The ViewCube comes with helpful features, as shown below:

- Each face of the cube gives you a 2D orthographic view, as discussed for the preset views. The upper corners of the cube are used to view an isometric view similar to SW, SE, NE, and NW. The lower corners of the cube will give you the same views but from below. In addition to these views, the ViewCube allows you to look at the model from the edges of the cube; these include 12 views. In total, there are 26 views.

- There is also a compass that allows you to rotate the ViewCube CCW or CW.

- AutoCAD saves one of the views as your Home view, and you can retain it by clicking the Home icon:

- Views are always related to the WCS (World Coordinate System) or to the UCS (User Coordinate System), so make sure you are at the right WCS or UCS (this topic will be discussed at the end of this chapter). Click the pop-up list at the bottom of the ViewCube to see something like the following:

- Also, at the bottom right of the ViewCube, you will also see a small triangle that contains a menu to help you change some settings. See the following illustration:

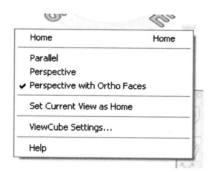

- You can do all or any of the following:
  - Retain the Home view.
  - Switch between Parallel (60/30 isometric) view, Perspective view, or Perspective view with Ortho Faces.
  - Change the Home view to be the current view.
  - Modify the ViewCube Settings.
  - Display the ViewCube Help topics.

- If you select **ViewCube Settings**, you will see the following dialog box:

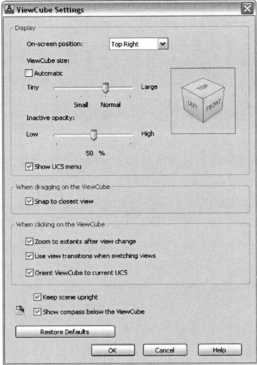

- Here, you can set the following:
  - The location of the ViewCube (default is Top Right).
  - The ViewCube size (default is Automatic).
  - The Inactive Opacity (the default is 50%).
  - To show the UCS menu or not.
- The rest of the settings are self-explanatory.

AUTOCAD 3D BASICS

 Practice 1-1

1. Start AutoCAD 2013.
2. Open **Practice 1-1.dwg**.

3. Using the preset views and the ViewCube (try all options), view the 3D model from all different angles.
4. Save and close the file.

1.3 ORBITING IN AUTOCAD

- Orbiting is rotating your model dynamically using the mouse. There are three types of orbiting in AutoCAD:
  - Orbit
  - Free Orbit
  - Continuous Orbit
- You will find all three commands in the **View** tab. Locate the **Navigate** panel and click the **Orbit** button, like the following:

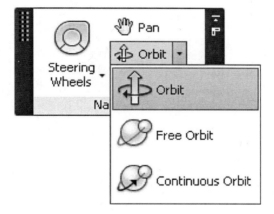

1.3.1 Orbit Command

- Once you start the command, the mouse cursor will change to:

- This command allows you to rotate your model around all axes. Your limit will be the Top view and the Bottom view, since you can't go beyond them. This command will end by pressing [Esc].
- *Another way to do the same thing is by holding [Shift]+click+mouse wheel.*

1.3.2 Free Orbit

- See the following illustration. An arcball will appear, showing four circles at each quadrant. Zoom and pan using your mouse wheel to center your model at the arcball:

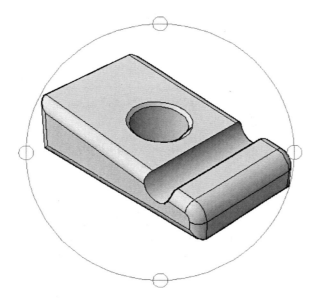

- Depending on where you are around the arcball, you will see a different cursor shape.
- You will see this shape ⊕ if you are at the circle at the right or left of the arcball. Click and drag the mouse to the right or left to rotate the shape around the vertical axis.
- You will see this shape ⊕ if you are at the circle at the top or bottom of the arcball. Click and drag the mouse up or down to rotate the shape around the horizontal axis.
- You will see this shape ⊕ if you are inside the arcball, which is similar to the Orbit command discussed above.
- You will see this shape ⊙ if you are outside the arcball, which allows you to rotate the model around an axis going through the center of the arcball and outside the screen. To end this command press [Esc].
- *Another way to access this command is by holding [Shift]+[Ctrl]+mouse wheel.*

1.3.3 Continuous Orbit

- When you issue this command, the mouse cursor will change to:

⊠

- This command allows you to begin non-stop orbiting of the model. All you have to do is to click and drag, and your model will start moving in the same direction you specified. To end this command, press [Esc].

1.4 STEERING WHEELS

- Steering Wheels allow you to navigate in the model using different methods such as zooming, orbiting, panning, walking, etc.
- To issue this command, go to the **View** tab, locate the **Navigate** panel, and click the **Steering Wheels** button, like the following:

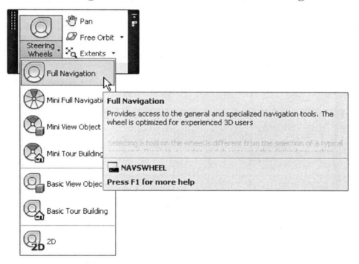

- Choose the **Full Navigation** option, and you will see the following:

- There are eight viewing commands; four located at the outer circle and four located at the inner circle. The outer circle includes:
  - Zoom command
  - Orbit command
  - Pan command
  - Rewind command
- In the inner circle, you will find:
  - Center command
  - Look command
  - Up/Down command
  - Walk command

1.4.1 Zoom Command

- This command allows you to zoom in on the 3D model (although using the mouse wheel will do almost the same job). Move your cursor to the desired location of your model, start the Zoom option, click and hold, and move your mouse forward to zoom in and move your mouse backward to zoom out.

1.4.2 Orbit Command

- This command is similar to the Orbit command. Move your cursor to the desired location, select the Orbit option, click and hold, and move the mouse right and left or up and down to orbit your model.

1.4.3 Pan Command

- This command allows you to pan in 3D. Move your cursor to the desired location, select the Pan option, click and hold, and move the mouse right and left or up and down to pan your model.

1.4.4 Rewind Command

- This command will record all the actions that took place using the other seven commands, as though you were rewinding a movie. Start the Rewind command, and you will see a series of screenshots; click and drag your mouse backwards to view all of your actions.

1.4.5 Center Command

- This command allows you to specify a new center point for the screen. (You should always hover over an object.) Select the Center option, click and hold, and then locate a new center point for the current view. When done, release the mouse, and the whole screen will move to capture the new center point.

1.4.6 Look Command

- This command assumes you are located somewhere and are not moving yet, but your head is moving up and down, left and right. Move the cursor to the desired location, select the Look option, click and hold, and move the mouse right and left or up and down.

1.4.7 Up/Down Command

- This command allows you to go up above the model or down below the model. Move your cursor to the desired location, then select the Up/Down option, and you will see a vertical scale; click and hold and move the mouse up and down.

1.4.8 Walk Command

- This command will simulate walking around; you have eight directions to choose from. If you combine this command with the Up/Down command, you will get the best results. Move your cursor to the desired location, select the Walk option, click and hold, and move the mouse in one of the eight directions.
- The other versions of the Steering Wheel in the **Navigate** panel in the **View** tab are mini copies of Full Navigation:
  - The Mini Full Navigation Wheel shows a small circle containing all the eight commands.
  - The Mini View Object Wheel shows a small circle containing four commands of the outer circle in the full wheel.
  - The Mini Tour Building Wheel shows a circle containing four commands of the inner circle in the full wheel.
  - The Basic View Object Wheel.
  - The Basic Tour Building Wheel.

1.5 VISUAL STYLES

- Visual styles determine how your objects are displayed. The default visual style in 3D templates is *Realistic*. To issue this command, go to the top-left side of the screen and click **Visual Style Controls**:

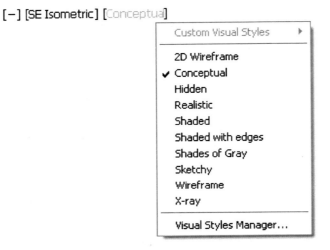

- Another way to reach the same is to go to the **Home** tab, locate the **View** panel, and then click the upper pop-up list:

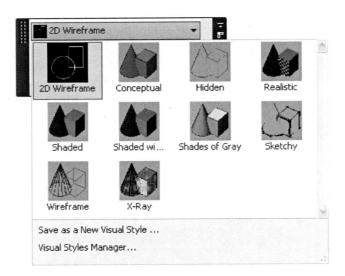

- AutoCAD comes with ten predefined visual styles (and you can make more):
  - With the 2D Wireframe visual style, you will see the edges of all objects. The lineweight and linetype of objects will be displayed.
  - With the 3D Wireframe visual style, you will see the edges of all objects without the lineweight and linetype.
  - With the 3D Hidden visual style, the view is similar to 3D Wireframe, but it will hide the obscured lines from the current viewpoint.
  - With the Conceptual visual style, objects will be shaded and smoothed.
  - With the Realistic visual style, objects will be shaded and smoothed; if materials are assigned, they will be shown with texture.
  - With the Shaded visual style, you will get the same result as Realistic but without texture.
  - With the Shaded with Edges visual style, you will get the same result as Shaded, except the edges will be highlighted.
  - With the Shades of Gray visual style, the colors of all objects will be shades of gray.
  - With the Sketchy visual style, you will see lines of edges extended beyond their limit with a jitter effect.
  - With the X-Ray visual style, objects will look transparent.
- The best way to learn about the different visual styles is to see them for yourself and experience the effects.
- Using the View panel you will see the effect of the visual style on the objects immediately whenever you select the desired visual style.

ORBIT, STEERING WHEEL, AND VISUAL STYLES

Practice 1-2

1. Start AutoCAD 2013.
2. Open **Practice 1-2.dwg**.
3. Using the Steering Wheel try to navigate through the high-rise buildings.
4. Using different orbit commands try to rotate the high-rise buildings to see them from different angles.
5. Using different visual styles, view the high-rise buildings with different styles.
6. Save and close the file.

1.6 WHAT IS THE USER COORDINATE SYSTEM (UCS)?

- In order to understand what the User Coordinate System (UCS) is we first have to discuss the World Coordinate System (WCS). There is one WCS in AutoCAD that defines all points in XYZ space, with XY as the ground and Z pointing upward. Since we need to build complex shapes in 3D, we need to work with all sorts of construction planes. That's why AutoCAD introduced the User Coordinate System, or UCS. When an AutoCAD user needs to work in a plane, AutoCAD gives him or her the power to create the desired UCS and create objects on it.
- The relationship between XYZ axes remains the same, and angles remain 90 degrees between one axis and another, according to the right-hand-rule. The following is the UCS icon:

- There are three ways to create a new UCS:
  - Manipulating the UCS icon
  - UCS command
  - DUCS

1.7 CREATING A NEW UCS BY MANIPULATING THE UCS ICON

- You can create a new UCS by manipulating the UCS icon. When you hover over the UCS icon it will turn gold. To activate the process, click the UCS icon, and it will change to the following:

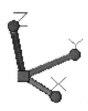

- As you can see, there is a grip at 0,0,0 and three spheres at the end of each axis. If you hover over the grip, it will show the following:

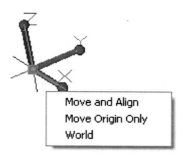

- There are three options to choose from:
  - Move and Align: This option allows you to move the UCS origin to a new location and align it with a face.
  - Move Origin Only: This option allows you to move the UCS origin to a new location.
  - World: This option allows you to retrieve the WCS.
- If you hover over one of the three spheres at the end of each axis, the following will be displayed:

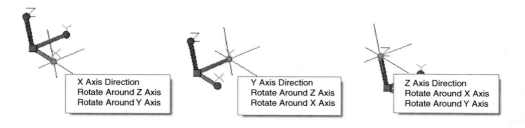

- You can do three things while in this menu:
  - Redirect the axis (X, Y, or Z) by specifying a point.
  - Rotate around the other two axes.
- Using all the methods, whether using the grip at 0,0,0 or the ones at the end of each axis, will help you to solve almost all UCS problems.

- See the following example.
- You want to move and align the UCS with the inclined face:

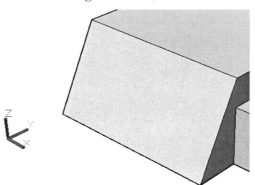

- Click the UCS icon, hover over the origin grip, and select the Move and Align option:

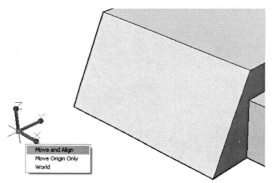

- Move the mouse to align with the inclined face. Once the UCS is located in the right position, click once, and you will get the following result:

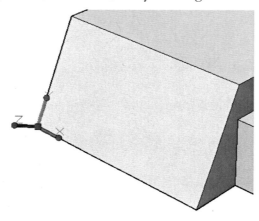

- Now you will be able to draw on this plane; see the following:

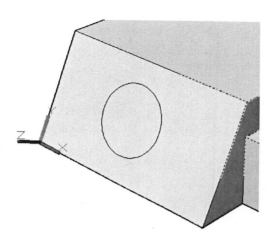

1.8 CREATING A NEW UCS USING THE UCS COMMAND

- This command will create a new UCS based on information supplied by you and doesn't ask for objects to be drawn before hand. This command offers so many ways to create a new UCS, which makes it a very important command for any AutoCAD user who wants to master 3D modeling in AutoCAD. We will discuss UCS options not alphabetically, but according to their importance. You can find all UCS options in the **Home** tab and the **Coordinates** panel.

1.8.1 3 Point Option

- To create any plane in XYZ space, you need three points. This fact makes this option the most practical option to create a new UCS. As the name of the option implies you will specify three points:
  - The first point is the new origin point.
  - The second point is any point on the positive X-axis.
  - The third point is a point on the positive Y-axis.

- To issue this command select the **3 Point** button:

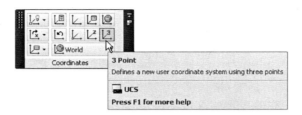

1.8.2 Origin Option

- This option will create a new UCS by moving the origin point from one point to another without modifying the orientation of the XYZ axes. To issue this command select the **Origin** button:

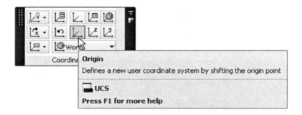

1.8.3 X/Y/Z Options

- These are three different options but they are very similar so we discuss them together. The idea is very simple; you fix one of the three axes and then you rotate the other two to get a new UCS. But how do you know whether the angle you have to input should be positive or negative? We will use the second right-hand-rule, which states: "Point the thumb of your right hand towards the positive portion of the fixed axis, and the curling of your four fingers will represent the positive direction." In the following illustration, we took X as an example. To issue the X command, select the **X** button:

1.8.4 Named Option

- If you will use a certain UCS a lot, you can save it and name it so you can retrieve it whenever you need it. After creating the UCS, go to the **View** tab, locate the **Coordinates** panel, and select the **Named** button:

- You will see the following dialog box:

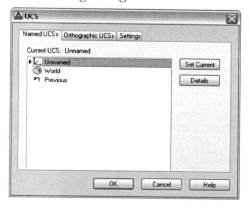

- Under the **Named UCSs** tab, you will see a list of the saved UCSs, plus *Unnamed*, *World*, and *Previous*. Unnamed is your current UCS, which you created before issuing this command. Click it to rename it, then type the new name and press [Enter], and you will see something like the following:

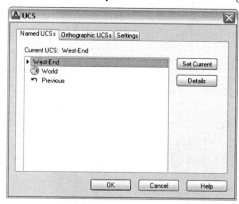

- To end the command, click **OK**.

- In the **Coordinates** panel there is a list of the preset UCSs, and you will see your saved UCS among them, as shown in the following:

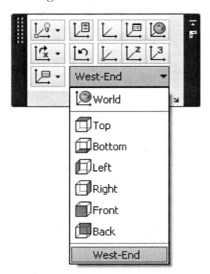

1.8.5 The Rest of the Options

- The above options were the most important ones, but we will go through the other options just to illustrate them:
  - The **World** option will restore the WCS:

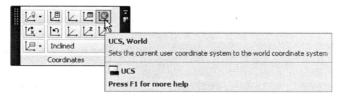

  - The **Z-Axis Vector** option will create a new UCS by specifying a new origin and a new Z-Axis:

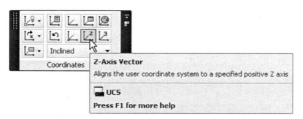

- The **View** option will create a new UCS parallel to the screen (this is good if you want to input text for the 3D model):

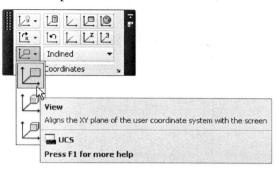

- The **Object** option will align the new UCS based on a selected object:

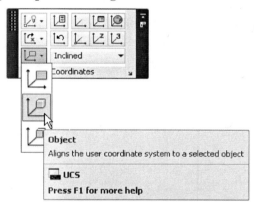

- The **Face** option will create a new UCS by selecting a face of a solid:

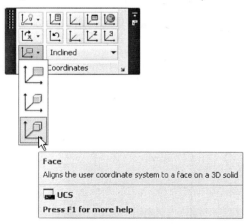

- The **Previous** option will restore the last UCS:

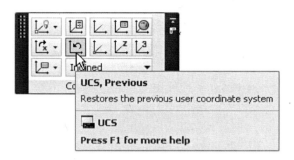

■ While you are hovering over the UCS icon, right-click and you will see the following menu:

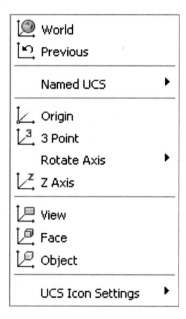

■ You will see all the options discussed above. This is another way to reach the same commands.

1.8.6 UCS Icon

■ The UCS icon is the XYZ symbol that resides at the lower-left corner of the screen in any file created by a 3D template. In the **Coordinates** panel, you can choose whether to show or hide the UCS icon and where. See the following:

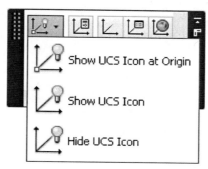

- This list allows you to:
  - **Show UCS Icon at Origin**; this option allows the UCS icon to move to the current origin. This option is very handy if you are in creating UCSs a lot.
  - **Show UCS Icon**; this option will always locate the UCS icon at the lower-left corner of the screen.
  - **Hide UCS Icon**; this option will not show the UCS icon on the screen (we don't recommend this option unless you know what you are doing).
- In the **Coordinates** panel, you will also find the **Properties** button, which allows you to control the shape, color, and size of the UCS icon:

- The following dialog box will be displayed:

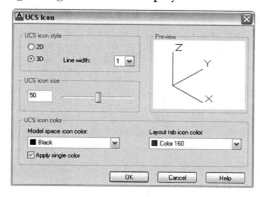

- In this dialog box you can change all or any of the following:
  - UCS icon style: 2D or 3D style of UCS icon and the line width.
  - Set the UCS icon size.
  - Set the UCS icon color.

1.9 DUCS COMMAND

- DUCS means Dynamic UCS. DUCS works only with other drafting and modifying commands. The procedure for creating a DUCS is simple and straightforward:
  - Activate DUCS on the status bar:

  - Start any drafting command (line, circle, box, cylinder, etc.).
  - To specify the first point, hover over the face you want to draw at, and when it highlights, AutoCAD has selected this face and you can specify the point and complete the command.
  - When you are done, the UCS prior to the command will be restored.
- DUCS has advantages and disadvantages. The advantages are it is fast and simple. But the disadvantages are:
  - It needs an existing object to be created.
  - It will be lost after finishing the drafting command.
  - It is limited to the faces of the selected object.
- Because of this, you should master all three UCS methods and utilize DUCS only when appropriate.

1.10 TWO FACTS ABOUT THE UCS AND DUCS

- There are two facts you should know about UCSs:
  - The ViewCube shows views in relation to the WCS or the current UCS, so we recommend that you always look at the ViewCube to understand

which UCS AutoCAD is work with to verify the accuracy of the view. See the following illustration:

Current UCS

- The second fact is that OSNAP has priority over UCS, which means you may choose points not in the current UCS using OSNAP, and OSNAP and UCS may contradict. You should be careful of this fact, and be extra cautious when defining a UCS and then choosing points not in the current UCS.

USER COORDINATE SYSTEM

Practice 1-3

1. Start AutoCAD 2013.
2. Open **Practice 1-3.dwg**.
3. Using any of the UCS creation methods, try to draw the following shapes on the existing shape to get this result:

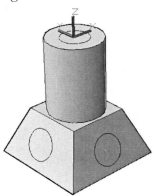

4. Save and close the file.

NOTES:

CHAPTER REVIEW

1. There are three Orbit commands.
   a. True
   b. False
2. _____ is a template to help you utilize the 3D environment.
3. Which of the following is *not* an option in the UCS command?
   a. Face
   b. Rotate around X
   c. Realistic
   d. 3 Point
4. Using the ViewCube, you can change the visual style.
   a. True
   b. False
5. Which one of the following is *not* an option of the Steering Wheel?
   a. Walk
   b. Center
   c. Conceptual
   d. Up/Down
6. _____ is a workspace that allows you to see all the Ribbons related to 3D.

CHAPTER REVIEW ANSWERS

1. a
3. c
5. c

Chapter 2 CREATING SOLIDS

In This Chapter
◇ Basic solid shapes
◇ Manipulating solids
◇ Press/Pull command
◇ 3D OSNAPs
◇ Subobjects and gizmos

2.1 INTRODUCTION TO SOLIDS

- Solids are 3D objects that contain faces, edges, vertices, and substance inside them. This makes them the most accurate 3D objects because they contain all the physical information.
- There are many ways to produce solids in AutoCAD; you can:
  - Create seven basic solid shapes.
  - Create complex solid shapes using Boolean operations (Union, Subtract, and Intersect).
  - Create complex solid shapes by converting 2D objects using the Press/Pull command.
  - Create complex solid shapes by using subobjects and gizmos.

2.2 CREATING SOLIDS USING BASIC SHAPES

- AutoCAD provides seven basic solid shapes. Basic shapes will be our first step to creating more complex shapes using other commands. To will find all of these commands got to the **Home** tab, locate the **Modeling** panel, and then select one of the following:

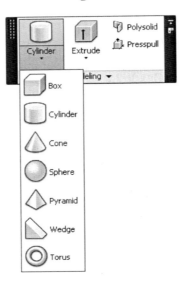

- Each command will be discussed in the following.

2.2.1 Box Command

- This command allows you to create a box or cube. Choose the **Box** command, and the following prompts will be displayed:

```
Specify first corner or [Center]:
Specify other corner or [Cube/Length]:
Specify height or [2Point]:
```

- To draw a box in AutoCAD, you have to define the base first, then the height. The base will be drawn on the current XY plane. Here are some variations:
  - Specify the base by typing the coordinates of the first corner and the coordinates of the opposite corner.
  - Specify the base by typing the coordinates of the first corner. To specify the opposite corner specify the length and height by typing and using the [Tab] key to jump.
  - Specify the base by specifying the first corner, then specifying length (in the X-axis direction) and width (in the Y-axis direction).

- Specify the base by specifying the first corner, then specifying the Cube option. AutoCAD will ask you to input one dimension for all the sides (length, width, and height).
- Specify the base by specifying the center point and one of the corners of the base.
- After specifying the base, specify the height, either by typing it in or by using the mouse or by specifying two points by clicking (this is a good method if you have a drawn object and the value of the height equals a distance on this object):

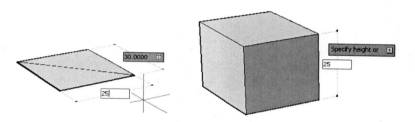

 ■ From now on, you can use Dynamic Input to input all or any of the distances needed to complete any command.

2.2.2 Cylinder Command

- This command allows you to create a cylinder. Choose the **Cylinder** command, and the following prompts will appear:

```
Specify center point of base or
[3P/2P/Ttr/Elliptical]:
Specify base radius or [Diameter]:
Specify height or [2Point/Axis endpoint]:
```

- To draw a cylinder in AutoCAD, you have to define the base first, then the height. Here are some variations:
  - The base could be a circle or ellipse. The options for drawing either a circle or ellipse are identical to the 2D prompts. The base will be drawn on the current XY plane.
  - After defining the base, specify the height, either by typing it in or by using the mouse or by using the 2 Points option (using the height of an existing object).

- The Axis endpoint option allows you to rotate the cylinder by 90° to make the height in the XY plane and not perpendicular to it. Using the Polar option you can direct the cylinder as you wish:

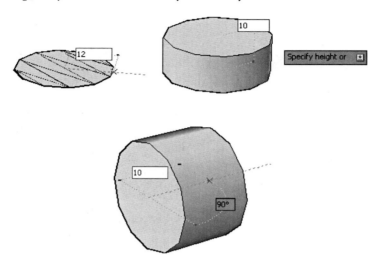

2.2.3 Cone Command

- This command allows you to create a cone. Choose the **Cone** command, and the following prompts will be displayed:

```
Specify center point of base or
[3P/2P/Ttr/Elliptical]:
Specify base radius or [Diameter]:
Specify height or [2Point/Axis endpoint/Top radius]:
```

- Cone and Cylinder have identical prompts, so we will not discuss them again, but the Cone command allows you to draw a cone using the top radius:

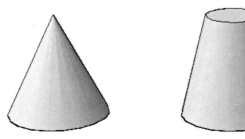

2.2.4 Sphere Command

- This command allows you to create a sphere. Choose the **Sphere** command, and the following prompts will appear:

```
Specify center point or [3P/2P/Ttr]:
Specify radius or [Diameter]:
```

- These prompts are identical to the Circle command prompts:

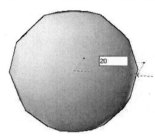

2.2.5 Pyramid Command

- This command allows you to create a pyramid. Choose the **Pyramid** command, and the following prompts will appear:

```
4 sides Circumscribed
Specify center point of base or [Edge/Sides]:
Specify base radius or [Inscribed]:
Specify height or [2Point/Axis endpoint/Top radius]:
```

- The default is a 4-sided base. You can change the number of sides by selecting the Sides option. To draw a base, use the same method for drawing a 2D polygon in AutoCAD. The Height options were discussed in the previous commands. The example below is for a 4-sided base, with the top a smaller polygon:

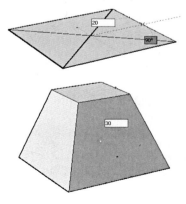

2.2.6 Wedge Command

- This command allows you to create a wedge. Choose the **Wedge** option, and the following prompts will be displayed:

```
Specify first corner or [Center]:
Specify other corner or [Cube/Length]:
Specify height or [2Point]:
```

- These prompts are identical to the Box prompts. In fact, a wedge is part of a box, as you have to define a base first then define the height. The wedge will be in the **YZ** plane towards the X-axis:

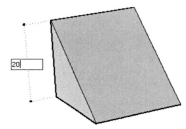

2.2.7 Torus Command

- This command allows you to create a torus. Choose the **Torus** option, and the following prompts will appear:

```
Specify center point or [3P/2P/Ttr]:
Specify radius or [Diameter]:
Specify tube radius or [2Point/Diameter]:
```

- Specify the center and radius (diameter) of the torus, and then specify the radius (diameter) of the tube. See the following illustration:

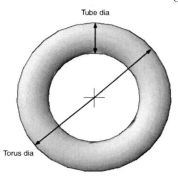

2.3 EDITING BASIC SOLID SHAPES USING GRIPS

- If you click one of the seven basic shapes, depending on the shape you will see grips in certain places. These grips allow you to modify the dimensions of the basic shape in all directions. Here are the box grips:

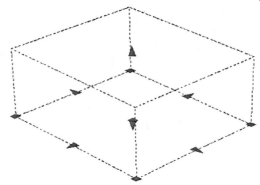

- Here are the cylinder grips:

- Here are the cone grips:

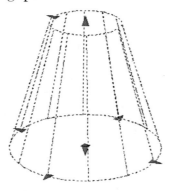

- Here are the sphere grips:

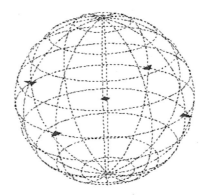

- Here are the pyramid grips:

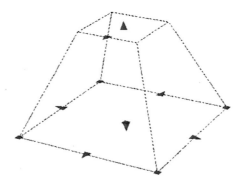

- Here are the wedge grips:

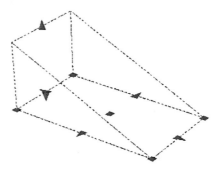

- Here are the torus grips:

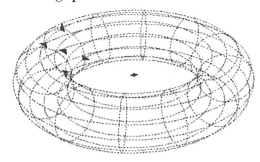

- In all of the above cases we can see a grip (rectangular shape) and an arrow (triangle shape). The grip will change two dimensions in one step; see the following illustration:

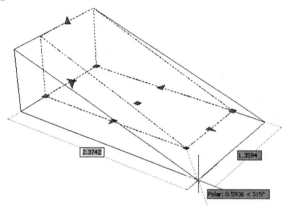

- Yet, the arrow will change one dimension in one direction, like the following:

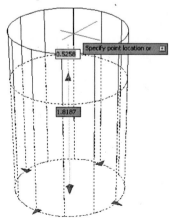

- You can type the input in both cases. If there is more than one dimension to enter, use the [Tab] key to jump from one number to another.
- You can use grips or arrows to see a dimension of the edges of the shapes. See the following illustrations:

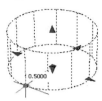 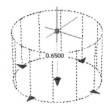

2.4 EDITING BASIC SOLID SHAPES USING QUICK PROPERTIES AND PROPERTIES

- While we can edit the basic shapes graphically using grips and arrows, we can also edit the dimensions of these basic shapes using the Quick Properties and Properties commands. When you click the basic solid shape once, the Quick Properties will come up automatically. See the following illustration:

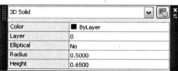

- These are the Quick Properties for the cylinder shape; you can change Radius and Height, and change the base from a circle to an ellipse. To activate Properties, select the desired shape, right-click, then select the **Properties** option, and you will see something like the following:

- At the top you will see **3D Solid**. Under **Geometry** you will see **Solid type** (in our case it is Cylinder, and you can't edit this piece of information) and all the geometry values, which can be edited and modified as desired.

2.5 3D OBJECT SNAP

- 3D Object Snap, or OSNAP, is similar to 2D Object Snap but only for 3D objects. This command will help you get a hold of points on 3D objects. As a first step, you have to switch **3D Object Snap** on/off using the status bar:

- Right-clicking the button will display the activated modes and deactivated modes, and you can switch any mode on/off:

- Selecting the **Settings** option will show the following dialog box:

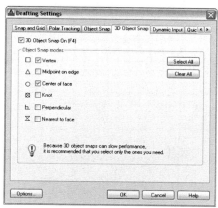

- You can turn on/off the desired modes. You have six modes to control:
  - You can snap to a vertex:

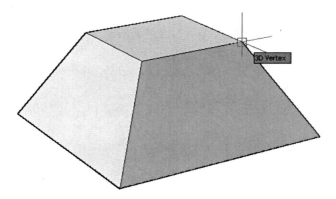

  - You can snap to a midpoint of an edge:

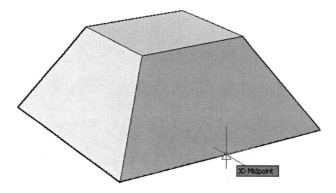

  - You can snap to the center of a face:

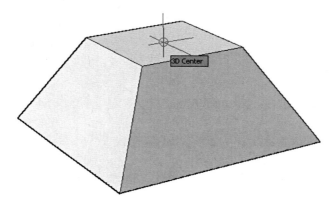

- You can snap to a knot of a spline:

- You can snap to a point perpendicular to a face.
- You can snap to a point on a face nearest to your selected point.

BASIC SOLID SHAPES AND 3D OSNAP

Practice 2-1

1. Start AutoCAD 2013.
2. Open **Practice 2-1.dwg**.
3. Create a pyramid with a 4-sided base (circumscribed), radius = 30, top side radius = 15, and height = 25.
4. Using the Cylinder command and 3D OSNAP, specify the base of the cylinder at the center of the top side of the pyramid using radius = 10 and height = 50.
5. Draw another cylinder at the top of the first one with radius = 2.5 and height = 75.
6. In an empty space, draw a box with length = 60, width = 2.5, and height = 30.
7. Using the Move command and 3D and 2D OSNAP, move the box to the upper cylinder to get the following result:

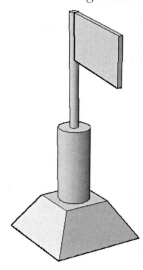

8. Using any of the editing methods discussed, change the radius of the top cylinder to = 5.
9. Change the width of the box = 5.
10. Save and close.

2.6 USING BOOLEAN FUNCTIONS

- This is our first method to create complex solid shapes. The three Boolean functions are Union, Subtract, and Intersect. To issue these commands, go to the **Home** tab, locate the **Solid Editing** panel, and you will see the following three buttons on the left:

2.6.1 Union Command

- This command allows you to union all selected solids or surfaces in a single object. You can choose your objects in any order you want, then press [Enter]. See the following example:

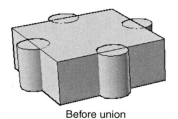   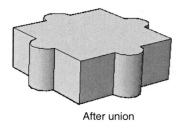

Before union   After union

2.6.2 Subtract Command

- This command allows you to deduct volumes of solids from an existing volume. When choosing objects, you should choose the objects to be deducted from first, then press [Enter], then choose the objects to be deducted, and then press [Enter] to end the command. See the following example:

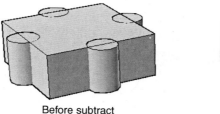
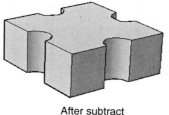

Before subtract   After subtract

2.6.3 Intersect Command

- This command will find and create the common volume between two or more solids. You can choose solids in any order you want. See the following example:

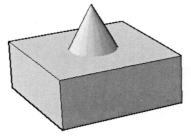

Before intersect   After intersect

2.6.4 Solid History

- The big question is "Does AutoCAD remember the original basic shapes used to create this complex solid?" The answer to this question is Yes!
- Take the following steps:
    - Before creating the complex solid shape go to the **Solid** tab, locate the **Primitive** panel, and click the **Solid History** button on (if it was not on already):

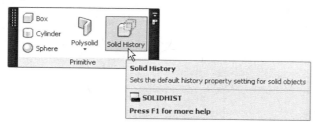

- Create your complex solid shape.
- Select the complex solid and activate the **Properties** palette, under **Solid History**, and make sure that **History = Record**. See the following illustration:

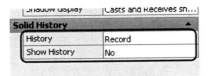

- Now hold the [Ctrl] key and hover over any part of your complex solid. The original objects will be highlighted, and accordingly, you can click to select them, which will enable the grips to appear, and you can edit their geometry values. See the following illustration:

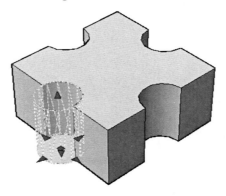

- Using **Properties**, if you set **Show History = Yes**, all solids used to create the complex solid will be displayed. But unfortunately you can't edit them in this state. To edit them you need to hold the [Ctrl] key as we did in the previous step. See the following illustration:

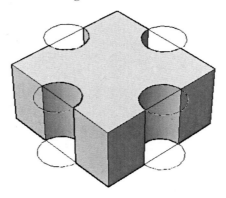

2.7 USING THE PRESS/PULL COMMAND

- The Press/Pull command is a very handy command that will help you create a 3D solid on the fly. You can do one of the following procedures:
  - Press/Pull can offset an existing face of a 3D solid, and depending on the mouse movement the volume of the 3D solid will increase or decrease.
  - After drawing a 2D shape on a face of a 3D solid, the Press/Pull command allows you to create a 3D solid, then union it or subtract it from the existing 3D solid.
  - Using a closed 2D shape, Press/Pull can create a 3D solid.
  - Using an opened 2D shape, Press/Pull can create a 3D surface (this will be discussed in Chapter 4).
- The Press/Pull command also has the **Multiple** option (you can also hold down the [Shift] key instead), which allows you to make several press/pull movements at the same time.
- To issue this command, go to the **Home** tab, locate the **Modeling** panel, and then select the **Press/Pull** button:

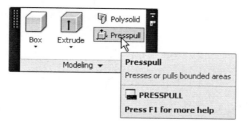

- You will see the following prompts:

```
Select object or bounded area:
Specify extrusion height or [Multiple]:
```

- The command will keep repeating these two prompts until you end it by pressing [Enter].
- Here are examples of the four cases:
  - Using the Press/Pull command to offset a 3D solid face:

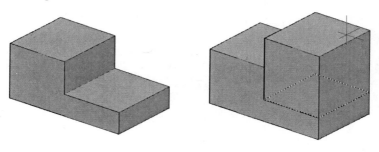

- After you draw a 2D object on a 3D solid face, you press or pull this 2D shape to add or remove volume from the total 3D solid:

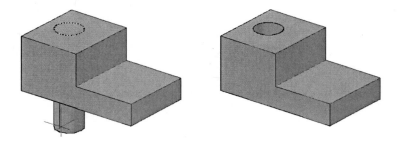

- You can press or pull a closed 2D object to produce a 3D solid:

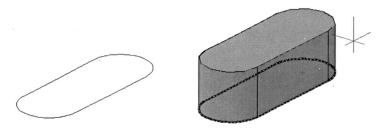

- You can press or pull an open 2D object to produce a 3D surface:

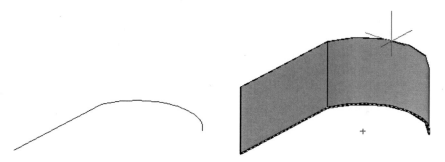

- Using the Multiple option (you can also hold the [Shift] key instead) allows you to select several objects (open 2D, closed 2D, face of a 3D solid) and press pull them in one step.

- You can also use the [Ctrl] key to control the output of the press/pull of a face of a solid. Take a look at the following example:

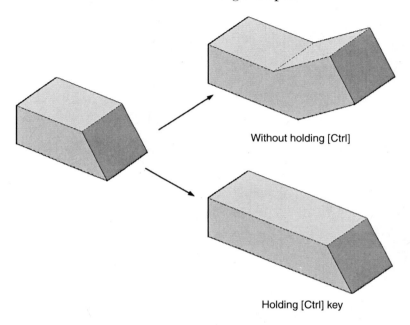

Without holding [Ctrl]

Holding [Ctrl] key

- In the upper image, you can see that Press/Pull offset the face perpendicular on the existing face. While holding the [Ctrl] key, Press/Pull offset it to follow the taper angle.

BOOLEAN FUNCTIONS AND PRESS/PULL

Practice 2-2

1. Start AutoCAD 2013.
2. Open **Practice 2-2.dwg**.
3. Select the big box, right-click, and select Properties. Change the value of History to Record. Close the Properties palette.
4. Subtract the upper cylinder from the box.
5. Subtract the lower cylinder from the box (use two different commands).
6. Union the four boxes at the edges with the big box.
7. Hold the [Ctrl] key and click the upper cylinder; when the grips show up, change the radius to 30 (i.e., add 10 units to the current radius).

8. Using the same technique, make the lower cylinder 10 units only (i.e., remove 10 units from the current radius).
9. Using the Press/Pull command pull the two upper sides up by 20 units.
10. Using the Press/Pull command press the front side to the inside by 15.
11. You will get something like the following:

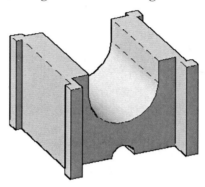

12. Thaw layer 2D Objects.
13. Using the Press/Pull command, pull the closed shape at the XY plane up by 60 units.
14. Using Press/Pull, press the four circles at the edge to remove them from the volume of the shape (you may need to use Subtract).
15. Using the Press/Pull command pull the two rectangles at the left side to the outside with distance = 15.
16. Using the Press/Pull command, pull the open shape up by 60 units.
17. Select the new shape. What type of shape is it? _____
18. This is the final shape:

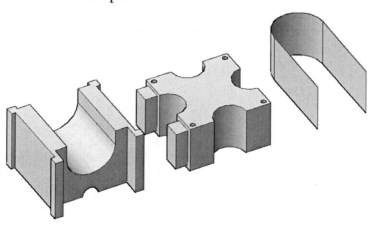

19. Save and close the file.

2.8 SUBOBJECTS AND GIZMOS

- A subobject is any part of a solid; it can be a vertex, an edge, or a face. A gizmo is a gadget that will help you Move, Rotate, or Scale the selected vertices, edges, or faces. Both are available in the **Home** tab and the **Selection** panel:

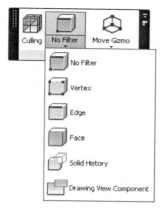 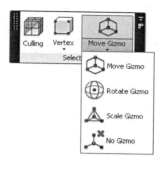

- The first step is to select the desired type of filter. Your cursor will change to the following shape (this cursor is for selecting vertices). Now when you select you will select only the subobject you chose:

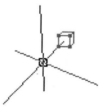

- The following are examples of selecting subobjects:

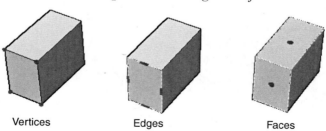

| Vertices | Edges | Faces |

- When you are selecting there are three things you have to consider; they are discussed as follows.

2.8.1 Culling

- The Culling feature is a very useful tool that allows you to select only sub-objects visible from the current viewpoint. To switch it on or off, go to the **Home** tab, locate the **Selection** panel, and then click the **Culling** button on or off:

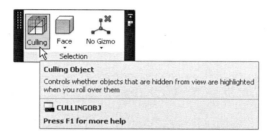

- If the Culling button is off, all subobjects will be selected no matter if they are visible from the current viewpoint or not. See the following two examples:

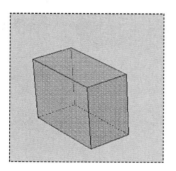

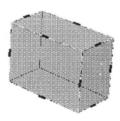

Culling is ON

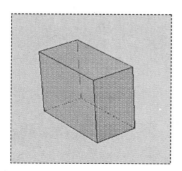

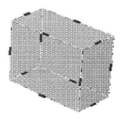

Culling is OFF

2.8.2 Selection Cycling

- When you click on a face (for example) to select it, you may have other faces behind it, and AutoCAD allows you to choose what exactly you intend to select. AutoCAD will show you a small window called **Selection,** which includes all the objects in the space you clicked. This small window will change from one object to another until you click the desired one. This feature is called **Selection Cycling**. To activate it go to the status bar and click the **Selecting Cycling** button:

- Here is an example. Let's assume we have a pyramid, and we clicked the top face as shown:

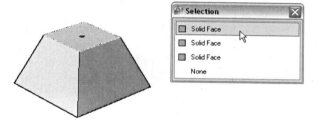

- AutoCAD will select the face nearest to your click, but will also allow you to select one of the two faces behind it as well. See the following illustration:

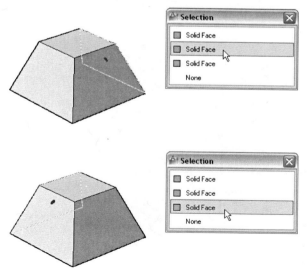

- This applies to edges as well.

2.8.3 Selecting Using Preset Views

- Using one of the Preset views will enable you to select multiple subobjects quickly and accurately. The following steps will clarify the concept:
    - Click **Culling** off.
    - Using the ViewCube, go to **Top** view.
    - From the Subobjects panel, select **Edge**.
    - Without issuing any command, use **Window** to contain all edges at the right side and at the left side of the box, just like the following:

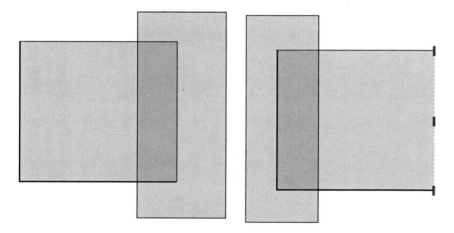

- Using the ViewCube, change to any 3D preset view, and you will see something like the following:

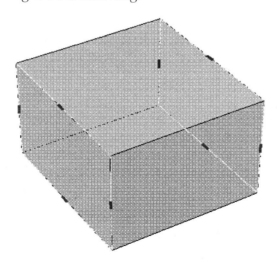

2.8.4 Using Gizmos

- After selecting the desired subobjects, you can choose the needed gizmo to change the subobjects. There are three gizmos to select from:
  - Move Gizmo:

  - Rotate Gizmo:

  - Scale Gizmo:

- By default, the gizmo will go to the last selected subobject. To move the gizmo from the last selected subobject to another, move your cursor (without clicking) to the desired subobject, stay for couple of seconds, and the gizmo will follow. You will see the following menu, if you right-click:

- This menu allows you to:
  - Change the Gizmo command.
  - Set constraints (this will set the axis or plane to be used).
  - Relocate, Align, and Customize Gizmo.
- The Move gizmo consists of the three axes to move the selected subobjects along one of them (i.e., X Y Z), or along a plane of two of them (i.e., XY, YZ, XZ). To do this, hover over one of the three axes until it highlights (it will become golden), then click and move. You can use the right-click menu to set the desired axis using the **Constraint** option. See the following illustration:

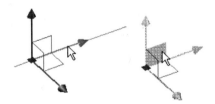

- The Rotate gizmo consists of three 3D circles. To rotate around one of the three axes, move your cursor to the desired circle and pause for a couple of seconds until it highlights (it will become golden), then click and rotate. You can use the right-click menu to set the desired axis using the **Constraint** option. See the following illustration:

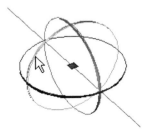

- The Scale gizmo consists of three axes. To scale you have to select a plane (XY, YZ, or XZ), or space (XYZ), and move your cursor to the desired plane and pause for couple of seconds until the plane highlights (it will become golden) and click, then scale. You can use the right-click menu to set the desired axis using the **Constraint** option. See the following illustration:

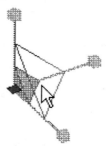

- Gizmos will not only act on subobjects, but also act on basic solid shapes. Take the following steps:
  - Make sure that Subobject = No Filter.
  - Make sure Culling is off (so you can see all grips).
  - When you click a basic solid shape, the gizmo will appear at the center of the shape; this will be your base point. If you want to use other base points, relocate the gizmo and select another grip as a new base point.
  - Start the desired command and execute it.
- See the following illustrations:

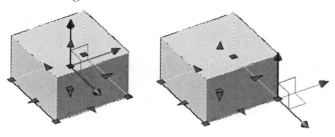

SUBOBJECTS AND GIZMOS

Practice 2-3

1. Start AutoCAD 2013.
2. Open **Practice 2-3.dwg**.
3. Change Subobjects to Face and move the upper face 5 units up.

4. Using the same face, rotate it around X (the red circle) 15°.
5. Select the edges as shown and move them to the outside 5 units:

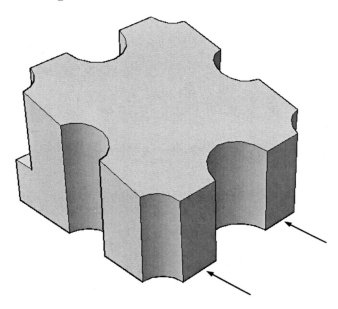

6. Rotate the shape to see the curved face at the back.
7. Select the curved face and scale it by a 1.5 factor.
8. Select the two vertices as shown and move them downwards by 5 units:

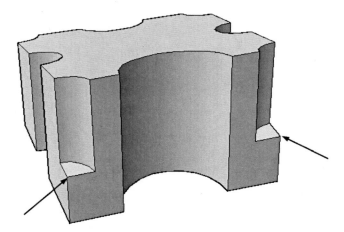

9. Save and close the file.

PROJECT USING METRIC UNITS

Project 2-1-M

1. Start AutoCAD 2013.
2. Open **Project 1-1-M.dwg**.
3. Draw the 3D solid shape using the following information:

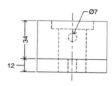

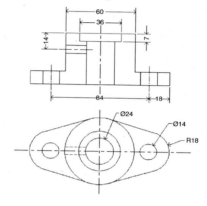

4. Save and close the file.

PROJECT USING IMPERIAL UNITS

Project 2-1-I

1. Start AutoCAD 2013.
2. Open **Project 1-1-I.dwg**.
3. Draw the 3D solid shape using the following information:

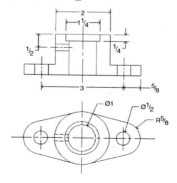

4. Save and close the file.

NOTES:

CHAPTER REVIEW

1. Boolean operations include three commands.
   a. True
   b. False
2. When you issue the Press/Pull command, you can press/pull any number of 2D objects in the same command.
   a. True
   b. False
3. Vertices, edges, and faces are _____ of a solid.
4. Which one of the following statements is *not* true?
   a. Solid History should be ON before the creation of a complex solid shape if you want to edit individual solid objects.
   b. If Culling = On, you can select all edges that are seen or not seen from the current viewpoint.
   c. Using preset views will help you select subobjects faster.
   d. There are three gizmo commands.
5. You can move vertices, edges and faces parallel to an axis or in a plane.
   a. True
   b. False
6. 3D Center is _____.
7. Which AutoCAD feature will help you select a face behind the face you clicked?
   a. Culling
   b. Selecting using Window
   c. Selection Cycling
   d. 3D OSNAP
8. Selection cycling and 3D OSNAP can be controlled from:
   a. Solid tab, Selection panel
   b. Home tab, Selection panel
   c. Status bar
   d. ViewCube

CHAPTER REVIEW ANSWERS

1. a
3. Subobjects
5. a
7. c

Chapter **3** CREATING MESHES

In This Chapter

◇ Basic mesh shapes
◇ Mesh subobjects and gizmos
◇ Creating meshes from 2D objects
◇ Converting, smoothing, refining, and creasing
◇ Face editing commands

3.1 WHAT ARE MESHES?

- The first time meshes as 3D objects were introduced was in AutoCAD 2010. Meshes look like solids except meshes don't have mass or volume. On the other hand, meshes have a unique capability that solids don't; they can be formed as different irregular shapes, which Autodesk calls *Free Form* design.
- Free Form is possible because meshes consist of faces (which you can specify), and faces consist of facets. You can increase the smoothness of a mesh by asking AutoCAD to take it to the next level, and AutoCAD will increase the number of facets to create a smoother mesh.
- This means facets are controlled by AutoCAD, and you can't edit them.

3.2 BASIC MESH SHAPES

- There are seven basic mesh shapes, just like solids. All the commands can be found in the **Mesh** tab and the **Primitives** panel:

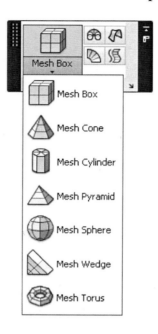

- Drawing a basic mesh shape is identical to drawing a basic solid shape, so your experience can be transferred to meshes easily. See the following illustration to differentiate between a solid and a box:

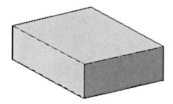 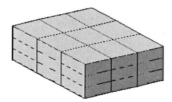

- To show the internal edges, set the system variable **vsedges** to 1.
- But how do you control the number of faces in each direction? To answer this question, you have to know that AutoCAD comes with default tessellation divisions. These values can be changed by going to the **Mesh** tab

and locating the **Primitives** panel, and then clicking the **Mesh Primitive Options** dialog box:

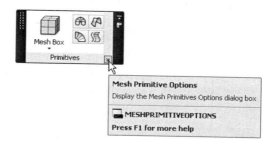

- You will see the following dialog box:

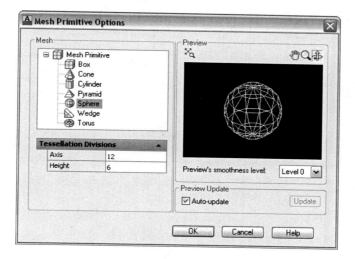

- You should choose the shape you want to alter, and then set the tessellation (number of faces) in all directions depending on the shape.

3.3 MANIPULATING MESHES USING SUBOBJECTS AND GIZMOS

- The fact that meshes have more subobjects than solids speaks volumes. Subobjects and gizmos give you the ability to manipulate any basic mesh shape to create an irregular shape using the three gizmo commands.

- See the following example:

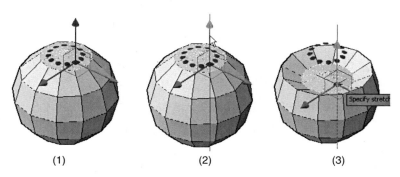

(1)   (2)   (3)

3.4 INCREASE/DECREASE MESH SMOOTHNESS

- Meshes can be created using four levels of smoothness: the lowest is None (no smoothness) and the highest is level 4. You can use Quick Properties to increase and decrease the smoothness of a mesh(es). See the following illustration:

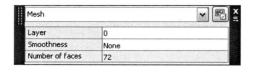

- Or you can go to the **Mesh** tab, locate the **Mesh** panel, and then select the **Smooth More** button or the **Smooth Less** button:

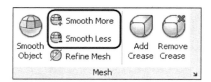

- AutoCAD will display the following prompts (for Smooth More):

```
Select mesh objects to increase the smoothness level:
Select mesh objects to increase the smoothness level:
```

■ See the following illustration:

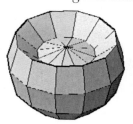
Smoothness = none

Smoothness = Level 2

BASIC MESH OBJECTS, SUBOBJECTS, AND GIZMOS

Practice 3-1

1. Start AutoCAD 2013.
2. Open **Practice 3-1.dwg**.
3. Go to the Mesh Primitive Options, select Sphere, and make sure that Axis = 12 and Height = 6.
4. Change the visual style to Conceptual (VSEDGES = 1).
5. Draw a mesh sphere with radius = 20.
6. Change subobject = Face.
7. Select all the top faces and move them 15 units down.
8. Select all the bottom faces and move them 15 units up.
9. Change subobject = No Filter.
10. Select the mesh and change the Smoothness = Level 2.
11. Change subobject = Edge.
12. Select the following edges:

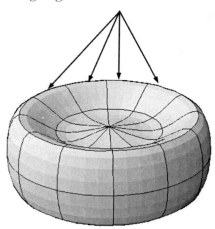

13. Move them upwards by 15 units.
14. You should have the following shape:

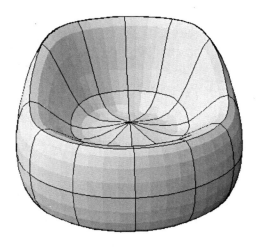

15. Save and close the file.

3.5 CONVERTING 2D OBJECTS TO MESHES

- AutoCAD is equipped with four commands to convert 2D closed or open shapes to meshes. These commands are:
  - Revolved Surface
  - Edge Surface
  - Ruled Surface
  - Tabulated Surface
- To set the density of the meshes generated from the above four commands, use the two system variables SURFTAB1 and SURFTAB2. The Ruled Surface and Tabulated Surface commands use only SURFTAB1, but the Revolved Surface and Edge Surface commands are affected by both SURFTAB1 and SURFTAB2.

3.5.1 Revolved Surface Command

- This command allows you to create a mesh from an open or closed 2D object by revolving it around an axis using an angle. To issue this command, go to the **Mesh** tab, locate the **Primitives** panel, and select the **Revolved Surface** button:

- You will see the following prompts:

```
Current wire frame density: SURFTAB1=16 SURFTAB2=16
Select object to revolve:
Select object that defines the axis of revolution:
Specify start angle <0>:
Specify included angle (+=ccw, -=cw) <360>:
```

- The first line reports the current values of SURFTAB1 and SURFTAB2, and this will be repeated with the other three commands.
- The second prompt asks you to select only one 2D open/closed object. Then you should select an object representing the axis of revolution. The fourth is used to specify the start angle if it is not 0 (zero). Finally, you should specify the included angle, keeping in mind that CCW is positive.
- The result will be something like this:

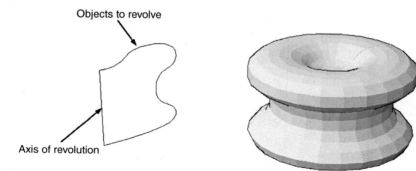

3.5.2 Edge Surface Command

- This command allows you to create a mesh from four 2D open shapes. For this command to be successful, the four objects should touch each other to form a closed four edged shape. To issue this command, go to the **Mesh** tab, locate the **Primitives** panel, and select the **Edge Surface** button:

- The following prompts will be displayed:

```
Current wire frame density: SURFTAB1=16 SURFTAB2=16
Select object 1 for surface edge:
Select object 2 for surface edge:
Select object 3 for surface edge:
Select object 4 for surface edge:
```

- The first line reports the current values of SURFTAB1 and SURFTAB2. Select the four edges one by one, and you will get something like the following:

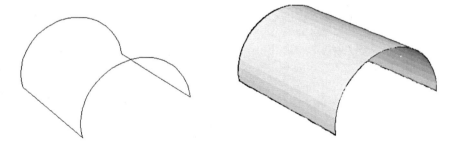

3.5.3 Ruled Surface Command

- This command will enable you to create a mesh between two objects, with one of the following variations:
    - Two open 2D shapes
    - Two closed 2D shapes
    - Open shape with point
    - Closed shape with point

- To issue this command, go to the **Mesh** tab, locate the **Primitives** panel, and select the **Ruled Surface** button:

- AutoCAD will display the following prompts:

```
Current wire frame density: SURFTAB1=24
Select first defining curve:
Select second defining curve:
```

- The first line reports the current value of SURFTAB1. The second and third prompts ask you to select the two defining curves open, closed, or point. See the following illustration:

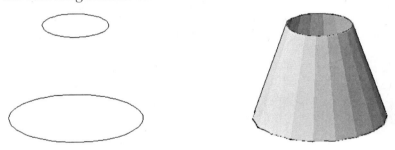

3.5.4 Tabulated Surface Command

- This command allows you to create a mesh using a 2D shape open or closed (path curve) and an object (direction vector). To issue this command, go to the **Mesh** tab, locate the **Primitives** panel, and select the **Tabulated Surface** button:

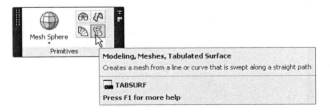

- AutoCAD will display the following prompts:

```
Current wire frame density: SURFTAB1=24
Select object for path curve:
Select object for direction vector:
```

- The first line reports the current value of SURFTAB1. The second asks you to select the path curve, and the third prompt asks you to select the direction vector. See the following illustration:

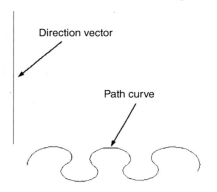

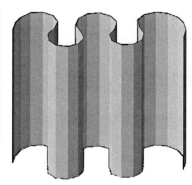

CONVERTING 2D OBJECTS TO MESHES

Practice 3-2

1. Start AutoCAD 2013.
2. Open **Practice 3-2.dwg**.
3. Set both SURFTAB1 and SURFTAB2 to = 24.
4. Change the visual style to Conceptual.
5. Using two arcs and the lines in the X direction, create an Edge surface.
6. Using the arcs and the lines in the Y direction, create a Rule surface.
7. Using the four circles and the vertical line, create a Tabulated surface.
8. Using the line and the polyline, create a Revolved surface.

9. You should have the following:

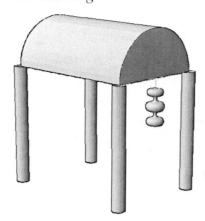

10. Save and close the file.

3.6 CONVERTING, REFINING, AND CREASING

- In this section we will discuss three commands that allow you to:
  - Convert legacy surfaces or solids to meshes.
  - Increase the number of faces in a mesh.
  - Add/remove creases to meshes.

3.6.1 Converting Legacy Surfaces or Solids to Meshes

- There were 3D objects called **surfaces** in AutoCAD before AutoCAD 2007. But these have nothing to do with the surfaces we will tackle in Chapter 4. If you open a file containing these types of objects and you want to convert them to meshes, or if you would like to convert any solid to a mesh, this command is for you. To issue this command, go to the **Mesh** tab, locate the **Mesh** panel, and then select the **Smooth Object** button:

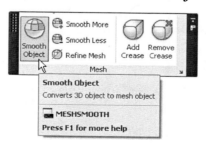

- The following prompts will be displayed:

```
Select objects to convert:
Select objects to convert:
```

- Select the desired 3D objects to convert, and when done press [Enter] to end the command. See the following illustration:

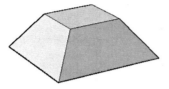
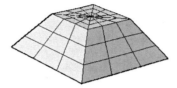

- But how can you control the result of the resultant mesh? How many faces will be in the different directions? Will it be smoothed or not? To answer these questions, go to the **Mesh** tab, locate the **Mesh** panel, and select the **Mesh Tessellation Options** button:

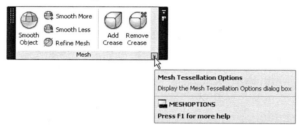

- The following dialog box will be displayed:

- If you select the **Mesh type** to be **Smooth Mesh Optimized,** then many of the options will be disabled. The rest of the options are self-explanatory.

3.6.2 Refining Meshes

- You can increase the number of faces in a mesh face, or in the whole mesh. The Smoothness level should be 1 or more. To issue this command, go to the **Mesh** tab, locate the **Mesh** panel, and then select the **Refine Mesh** button:

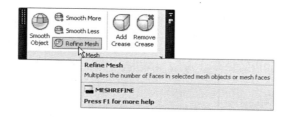

- The following prompts will be displayed:

```
Select mesh object or face subobjects to refine:
Select mesh object or face subobjects to refine:
```

- Select a mesh or a face subobject. You will see the following message if you try to refine a mesh with Smoothness level = None:

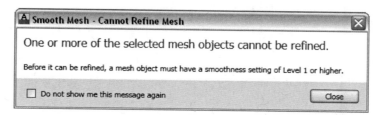

- When done press [Enter] to end the command. See the following illustration:

 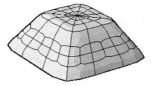

3.6.3 Add or Remove Crease

- This command is the opposite of smoothness, as you can add sharpness to some face subobjects of your mesh. To issue this command, go to the **Mesh** tab, locate the **Mesh** panel, and then select the **Add Crease** button or the **Remove Crease** button:

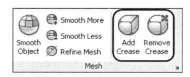

- You will see the following prompts:

```
Select mesh subobjects to crease:
Select mesh subobjects to crease:
Specify crease value [Always] <Always>:
```

- You can crease faces, edges, or vertices. The last prompt is used to specify the **crease value**, which is the highest Smoothness level value that will preserve the crease; above this value the crease will be smoothed. **Always** means the crease will always be preserved. See the following illustration:

 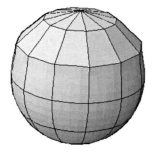

- If you click a mesh, two new context panels will be added to the *current* tab: the **Smooth** panel, which includes all the commands discussed above, and the **Convert Mesh**, which will be discussed later in the book.

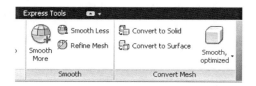

CONVERTING, REFINING, AND CREASING

 Practice 3-3

1. Start AutoCAD 2013.
2. Open **Practice 3-3.dwg**.
3. Change the Smoothness = level 1.
4. Refine the two faces as shown below:

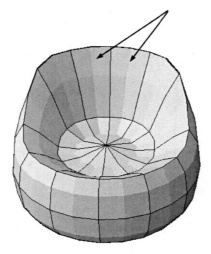

5. Add crease to the two faces behind the faces selected in the previous step.
6. Change the Smoothness to level 4.
7. Thaw layer Legs.
8. Convert the four spheres to meshes.
9. This is the final result:

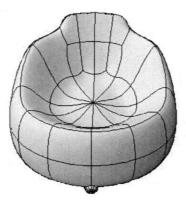

10. Save and close the file.

3.7 MESH EDITING COMMANDS

- Using a mesh you can do all or any of the following:
    - Split faces
    - Extrude faces
    - Merge faces
    - Close a hole in a mesh
    - Collapse a face or edge
    - Spin a triangular face

3.7.1 Split Mesh Face

- This command allows you to split a single face in a mesh into two faces using two points or two vertices. To issue this command, go to the **Mesh** tab, locate the **Mesh Edit** panel, and select the **Split Face** button:

- The following prompts will displayed:

```
Select a mesh face to split:
Specify first split point on face edge or [Vertex]:
Specify second split point on face edge or [Vertex]:
```

- The first prompt asks you to select the desired face. The cursor will change to the following:

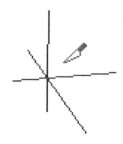

- Now you have to select the first split point, then the second split point. You can select any point using OSNAP or 3D OSNAP to choose points, or you can choose the **Vertex** option to choose vertices. See the following illustration:

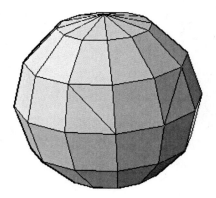

3.7.2 Extrude Face

- This command allows you to extrude a face perpendicular to the face plane either outside, or inside. To issue this command, go to the **Mesh** tab, locate the **Mesh Edit** panel, and then select the **Extrude Face** button:

- You will see the following prompts:

```
Adjacent extruded faces set to: Join
Select mesh face(s) to extrude or [Setting]:
Select mesh face(s) to extrude or [Setting]:
Specify height of extrusion or [Direction/Path/Taper angle]:
```

- Select the mesh face(s) to extrude, then specify the height of extrusion. The rest of the options will be discussed in Chapter 5. See the following illustration:

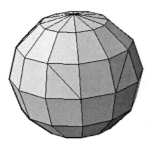 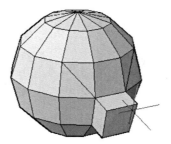

3.7.3 Merge Faces

- This command is the opposite of Split Face. It allows you to merge two or more adjacent mesh faces. To issue this command, go to the **Mesh** tab, locate the **Mesh Edit** panel, and then select the **Merge Face** button:

- The following prompts will be shown:

```
Select adjacent mesh faces to merge:
Select adjacent mesh faces to merge:
Select adjacent mesh faces to merge:
```

- Select the preferred faces to be merged, and then press [Enter]. See the following illustration:

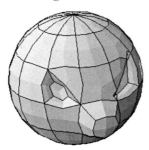 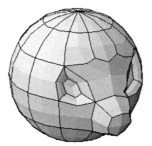

3.7.4 Closing a Hole in a Mesh

- There is no command to make a hole in a mesh, because it is simple to select face(s) and press the [Del] key. In order to close a hole in a mesh, go to the **Mesh** tab, locate the **Mesh Edit** panel, and select the **Close Hole** button:

- The following prompts will be shown:

```
Select connecting mesh edges to create a new mesh face:
Select connecting mesh edges to create a new mesh face:
```

- Start by selecting all the edges of the hole (they should be in the same plane), and when done, press [Enter] to end the command. See the following illustration:

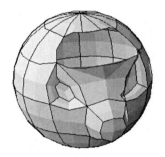 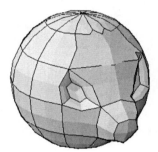

3.7.5 Collapse a Face or Edge

- This command is used to delete a face or edge without making a hole, but also to merge the vertices of selected faces or edges, as if you are reducing the number of faces in the mesh. To issue this command, go to the **Mesh** tab, locate the **Mesh Edit** panel, and select the **Collapse Face or Edges** button:

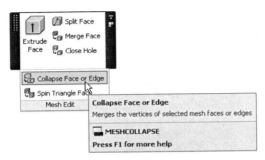

- The following prompts will be shown:

```
Select mesh face or edge to collapse:
```

- Select the desired edges or faces. See the following illustration:

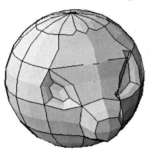 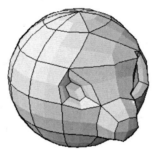

3.7.6 Spin Triangular Face

- This command allows you to spin the shared edge of two triangular mesh faces. To issue this command, go to the **Mesh** tab, locate the **Mesh Edit** panel, and select the **Spin Triangular Face** button:

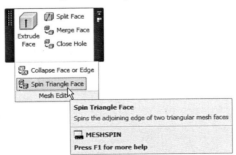

- The following prompts will be shown:

```
Select first triangular mesh face to spin:
Select second adjacent triangular mesh face to spin:
```

- Select the two triangular faces sharing the same edge, and you will see something like the following:

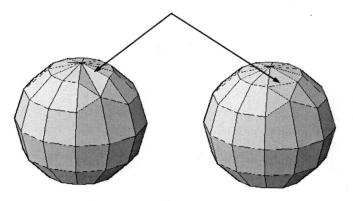

MESH EDITING COMMANDS

Practice 3-4

1. Start AutoCAD 2013.
2. Open **Practice 3-4.dwg**.
3. Split the two faces as shown below, and extrude the face by 7 units:

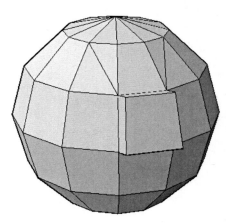

4. Split the face as shown below then extrude the faces by 2.5 units to the inside to get the following shape:

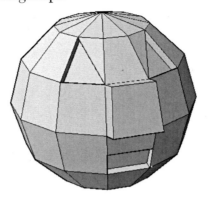

5. Collapse the edges shown below:

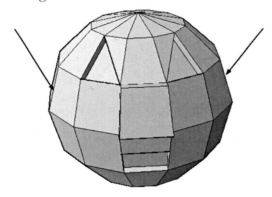

6. Change the Smoothness = Level 4.
7. You will get the following shape:

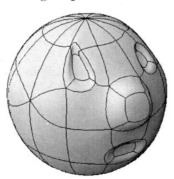

8. Save and close the file.

PROJECT USING METRIC UNITS

Project 3-1-M

1. Start AutoCAD 2013.
2. Open **Project 3-1-M.dwg**.
3. Change the visual style to Conceptual (change VSEDGES = 1).
4. Start the Mesh Primitives Options, and set the following values for the Box: Length = 8, Width = 8, Height = 1.
5. Draw a mesh box 120x120x60.
6. Select all the faces as shown below:

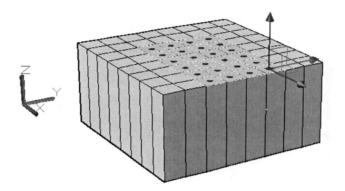

7. Using the gizmo to move the faces downwards by 32 units.
8. Select the edges as shown below:

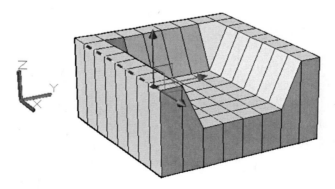

9. Move the edges towards the negative of the Y-axis by 20 units.
10. Move the same edges downward by 10 units.
11. Do the same for the other side.

12. Split faces from the midpoints and extrude faces to the outside by 12 units:

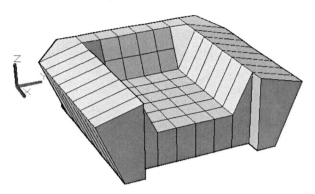

13. Merge the faces as shown below, then move it downwards by 6 units:

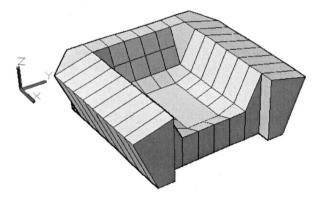

14. Move the faces upwards by 16 units as shown below:

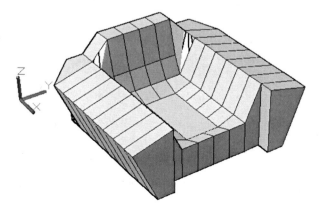

15. Change the Smoothness of the shape to level 4.
16. Select the faces as shown below and add crease by value = 1:

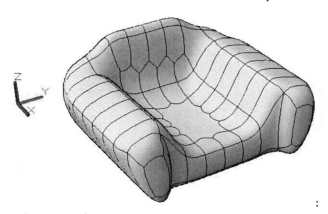

17. Save and close the file.

PROJECT USING IMPERIAL UNITS

Project 3-1-I

1. Start AutoCAD 2013.
2. Open **Project 3-1-I.dwg**.
3. Change the visual style to Conceptual (change VSEDGES = 1).
4. Start the Mesh Primitives Options, and set the following values for the Box: Length = 8, Width = 8, Height = 1.
5. Draw a mesh box 48"x48"x24".
6. Select all the faces as shown below:

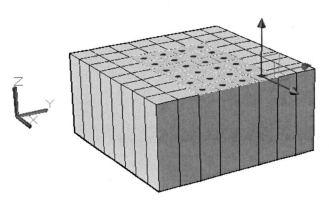

7. Using the gizmo move the faces downwards by 13".
8. Select the edges as shown below:

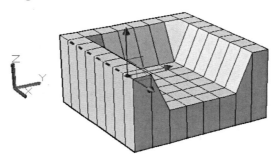

9. Move the edges towards the negative of the Y-axis by 8".
10. Move the same edges downward by 4".
11. Do the same for the other side.
12. Split faces from the midpoints and extrude faces to the outside by 5":

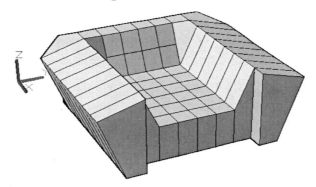

13. Merge the faces as shown below, then move it downwards by 3":

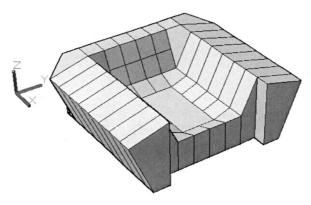

14. Move the faces upwards by 6" as shown below:

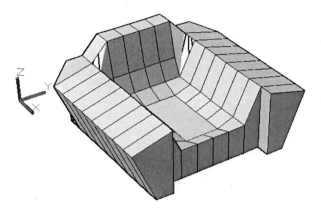

15. Change the Smoothness of the shape to level 4.
16. Select the faces as shown below and add crease by value = 1:

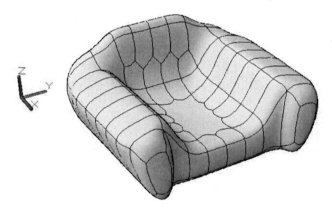

17. Save and close the file.

NOTES:

CHAPTER REVIEW

1. SURFTAB1 and SURFTAB2 will affect which of the following?
   a. Tabulated Surface and Ruled Surface
   b. Tabulated Surface, Revolved Surface, and Ruled Surface
   c. Tabulated Surface, Revolved Surface, Ruled Surface, and Edge Surface
   d. None of the above
2. There are three levels of smoothness for meshes.
   a. True
   b. False
3. Faces, edges, and vertices can be controlled by you, but you can't control _____ in meshes.
4. In order to see the internal edges of a mesh, you should change which of the following system variables?
   a. VSEDGECOLOR
   b. VSEDGELEX
   c. VSEDGES
   d. EDGEDISPLAY
5. In Revolved Surface, the axis of revolution could be any two points.
   a. True
   b. False
6. The _____ command will add sharpness to a mesh.
7. Ruled Surface can deal with:
   a. Closed 2D objects
   b. Points
   c. Open 2D objects
   d. All of the above
8. You can delete a face of a mesh by:
   a. Using the DELMESHFACE command
   b. Selecting the face and pressing [Del] on the keyboard
   c. Going to the Mesh tab, locating the Mesh Edit panel, and clicking the Delete button
   d. You can't delete a single mesh face

CHAPTER REVIEW ANSWERS

1. c
3. Facets
5. b
7. d

Chapter 4

CREATING SURFACES

In This Chapter

◇ Creating surfaces using Planar, Network, Blend, Patch, and Offset
◇ Editing surfaces
◇ NURBS and control vertices

4.1 INTRODUCTION TO SURFACES

- Before AutoCAD 2007, there were 3D objects called surfaces, but these objects are not related to what we are discussing here.
- In AutoCAD there are two types of surfaces:
  - Procedural Surfaces: This type of surface keeps a relationship with objects around it and will change if they change. Procedural Surfaces are associative.
  - NURBS Surfaces: This type of surface has Control Vertices (CVs) that allow you to manipulate the surface as if you are sculpting. NURBS Surfaces are not associative.
- In order to control what type of surface you are creating, go to the **Surface** tab, locate the **Create** panel, and pay attention to the two buttons at the right. If the **NURBS Creation** button is on, then the surface is NURBS. If **Surface Associativity** is on, then the surface is Procedural:

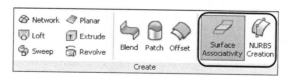

- This is true for all types of surfaces, except the Planer Surface.

4.2 CREATING A PLANAR SURFACE AND NETWORK SURFACE

- The first two commands to create surfaces are:
  - Planar Surface
  - Network

4.2.1 Planar Surface

- Using this command allows you to create rectangle or irregular planar surfaces on the current XY plane. To issue this command, go to the **Surface** tab, locate the **Create** panel, and then select the **Planar** button:

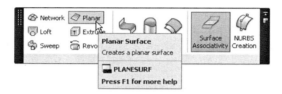

- The following prompts will be displayed:

```
Specify first corner or [Object] <Object>:
Specify other corner:
```

- Specify the first corner and the other corner to draw a rectangular shape surface. See the following illustration:

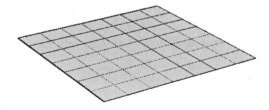

- On the other hand, you can convert any 2D shape to a planar surface using the following prompts:

```
Specify first corner or [Object] <Object>: O
Select objects:
Select objects:
```

- Select the desired object and you will get something like the following:

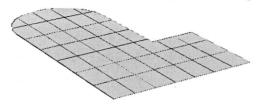

- To control how many lines will be shown inside the surface, use the two system variables SURFU and SURFV. You have to type these two commands at the Command window.

4.2.2 Network Surface

- This command allows you to create a surface from the network of curves in different UCSs, even if they are not connected. Go to the **Surface** tab, locate the **Create** panel, and then select the **Network** button:

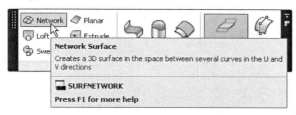

- The following prompts will be displayed:

```
Select curves or surface edges in first direction:
Select curves or surface edges in second direction:
```

- When you create the curves to be used in this command, you have to make sure to create them in two directions, U and V. Select curves in the U direction (can be more than one curve), then press [Enter], then select curves in the V direction, and then press [Enter] to see the result. See the following illustration:

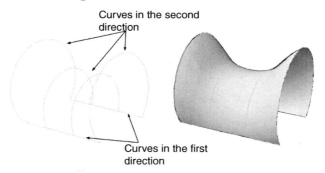

Curves in the second direction

Curves in the first direction

- The same system variables that control planar surfaces control network surfaces, SURFU and SURFV.

CREATING PLANAR AND NETWORK SURFACES

 Practice 4-1

1. Start AutoCAD 2013.
2. Open **Practice 4-1.dwg**.
3. The four cylinders are procedural surfaces. In order to create them we used circles at the top of each cylinder.
4. Select the circle at the top of one of them (make sure that Selection Cycling is on) and change the radius to 3. Notice how the whole surface changed accordingly.
5. Do the same for the other three surfaces.
6. Copy the circle at the top of one of the cylinders and create a planar procedural surface from it.
7. Copy the newly created surface to the top of the four cylinders.
8. Using the NURBS surface and the Network command, create the cover using the four arcs and the two lines.
9. You should get something like the following:

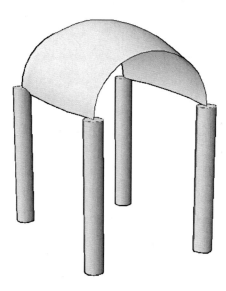

10. Save and close the file.

4.3 USING THE BLEND, PATCH, AND OFFSET COMMANDS

The Blend, Patch, and Offset commands allow you to create complex surfaces.

4.3.1 Blend Command

- This command allows you to create a continuity surface between existing surface edges. To issue this command, go to the **Surface** tab, locate the **Create** panel, and select the **Blend** button:

- The following prompts will be displayed:

```
Continuity = G1 - tangent, bulge magnitude = 0.5
Select first surface edges to blend or [Chain]:
Select first surface edges to blend or [Chain]:
Select second surface edges to blend or [Chain]:
Select second surface edges to blend or [Chain]:
```

- The first line reports the current values for both **Continuity** and **Bulge magnitude**. AutoCAD asks you to select the edges in two sets; once you finish the first set, press [Enter], and start selecting the second set of edges, and when done press [Enter]. Use the **Chain** option to select all edges adjunct to each other. At the end you will see the following prompts:

```
Press Enter to accept the blend surface or [CONtinuity/Bulge
magnitude]:
```

- Press [Enter] to accept the default values, or you can change both values to something else. You will see the following:

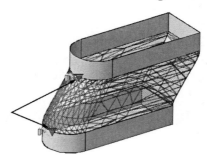

- The two triangles allow you to change the value of continuity at the two edges selected:

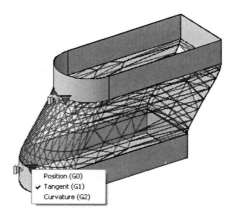

- The continuity values are:
  - G0 – Position
  - G1 – Tangent
  - G2 – Curvature
- G0 is a flat surface, G1 will be more curved, and both make sure that the new surface is tangent to the two edges selected. G2 is the best curvature. You can also change the value of Bulge Magnitude, which is like the swelling value; the default value is 0.5, the lowest is 0, and the highest is 1. This is the final result:

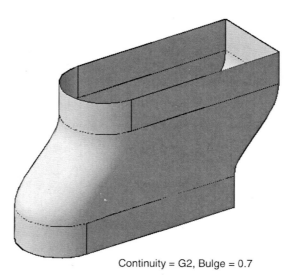

Continuity = G2, Bulge = 0.7

4.3.2 Patch Command

- This command will create a surface to connect the selected edges of an existing surface. To issue this command, go to the **Surface** tab, locate the **Create** panel, and select the **Patch** button:

- The following prompts will be displayed:

```
Continuity = G0 - position, bulge magnitude = 0.5
Select surface edges to patch or [CHain/CUrves] <CUrves>:
```

- The first line reports the current values for both continuity and bulge magnitude. AutoCAD then asks you to select surface edges to be patched; when done press [Enter]. You can use the Chain option to select all adjacent edges in one shot. Accordingly the following prompts will be shown:

```
Press Enter to accept the patch surface or [CONtinuity/Bulge
magnitude/Guides]:
```

- Press [Enter] to accept the default values, or you can change both values to new values. The other option is **Guides**; the following prompt will be shown:

```
Select curves or points to constrain patch surface:
```

- You need to select whether to constrain the patching process. See the following two illustrations. The first one is without guides:

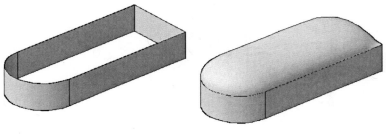

Continuity = G1

- The second one is with guides:

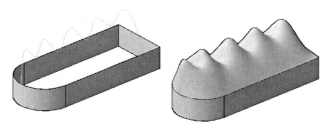

4.3.3 Offset Command

- This command will help you create a parallel copy of an existing surface. To issue this command, go to the **Surface** tab, locate the **Create** panel, and then select the **Offset** button:

- The following prompts will be shown:

```
Connect adjacent edges = No
Select surfaces or regions to offset:
```

- The first line gives you the current value for **Connect adjacent edges**. AutoCAD will then ask you to select the desired surface(s); when done, press [Enter].
- By selecting a surface, you will see something similar to the following:

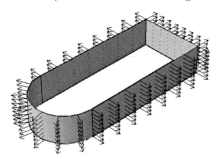

- By that time, the following prompts will be shown:

```
Specify offset distance or [Flip direction/Both sides/Solid
/Connect] <0.0000>:
```

- The above illustration shows the default offset direction, but you can:
  - Use the **Flip direction** option to flip the direction of the offset.
  - Use the **Both sides** option to specify the offset to both sides.
  - Use the **Solid** option to create a new solid object that consists of the original surface and the parallel surface to be produced.
  - Use the **Connect** option to connect multiple offset surfaces if the original surfaces were connected.

BLEND, PATCH, AND OFFSET COMMANDS

Practice 4-2

1. Start AutoCAD 2013.
2. Open **Practice 4-2.dwg**.
3. Offset the upper shape to the outside using value 0.2 and make sure to connect the resultant surface together.
4. Do the same thing to the lower shape but to the inside.
5. Using the Blend command, blend the edges of the upper inner shape with the lower inner shape using the Chain option, then specify continuity to be G2 and bulge magnitude = 0.75.
6. Do the same thing for the two outer shapes.
7. Using the Patch command, cover the lower part of the shape using the default values for continuity and bulge.
8. You should get the following shape:

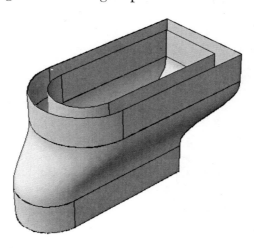

9. Save and close the file.

4.4 HOW TO EDIT SURFACES

- AutoCAD has special editing commands for surfaces, and three of the following work about the same way they do in 2D, but the last one allows you to convert a surface to a solid. These commands are:
  - Fillet
  - Trim and Untrim
  - Extend
  - Sculpt

4.4.1 Fillet Command

- This command is similar to the 2D version; it will create a curved surface from selected edges. You have to select two surfaces just as we did in 2D. This command will work only on surfaces. To issue this command, go to the **Surface** tab, locate the **Edit** tab, and select **Fillet** button:

- The following prompts will be shown:

```
Radius = 1.0000, Trim Surface = yes
Select first surface or region to fillet or [Radius/Trim surface]:
Select second surface or region to fillet or [Radius/Trim surface]:
Press Enter to accept the fillet surface or [Radius/Trim surfaces]:
```

- The first line reports the current radius value and the current surface trimming settings. After reading the first line, select the first surface, then select the second surface. If you press [Enter] you will accept the default values for radius and trimming. You have the ability to change the radius either by typing or using the mouse. See the shape below:

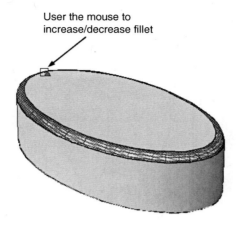

User the mouse to
increase/decrease fillet

4.4.2 Trim Command

- This command will help you trim surfaces (make openings) using a 2D object or another surface. To issue this command, go to the **Surface** tab, locate the **Edit** tab, and select the **Trim** button:

- You will see the following prompts:

```
Extend surfaces = Yes, Projection = Automatic
Select surfaces or regions to trim or [Extend/PROjection direction]:
Select cutting curves, surfaces or regions:
Select cutting curves, surfaces or regions:
Select area to trim [Undo]:
Select area to trim [Undo]:
```

- The first line gives you the current settings for **Extending surfaces** and **Projection**. Then AutoCAD asks you to select the surface(s) to be trimmed; when done press [Enter]. In the next step, AutoCAD asks you to select the cutting curves, which can be any 2D object or surface. Then select the area to trim by clicking on the area (part) that you want to remove.

- **Option Extend** is ideal when a surface will be trimmed based on another surface, and the cutting surface does not intersect with the surface to be trimmed. If this option is set to yes, AutoCAD will always trim, but if it is set to no, then the trimming operation will not take place, except when the cutting surface intersects the surfaces to be trimmed:

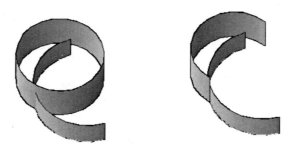

Extension = yes

- The Projection option is ideal when you have a 2D object as your cutting curve, and it is not fully in the range of surfaces to be trimmed. The question here is whether AutoCAD should project the 2D object on the surface or not? See the following illustration:

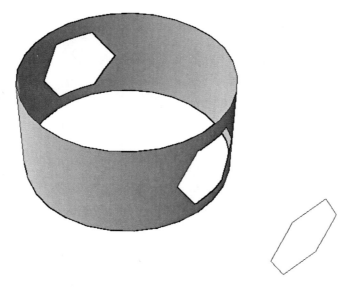

- If the surface was Associative, if you erase the 2D object, the two openings will be deleted as well. Also, any change on the 2D object will be reflected on the openings.

4.4.3 Untrim Command

- This command is the opposite of the Trim command; it will cover the areas trimmed by the Trim command. To issue this command, go to the **Surface** tab, locate the **Edit** tab, and select the **Untrim** button:

- The following prompts will be displayed:

```
Select edges on surface to un-trim or [SURface]:
Select edges on surface to un-trim or [SURface]:
```

- There are two ways to untrim; either select the edges of the gap you want to close or select option **SURface**, which means you will select the whole surface that contains the gap and AutoCAD will close all the gaps in it. See the following:

Before Untrim   After Untrim

4.4.4 Extend Command

- This command will increase the length of the selected surface's edges. To issue this command, go to the **Surface** tab, locate the **Edit** tab, and select the **Extend** button:

- The following prompts will be shown:

```
Modes = Extend, Creation = Append
Select surface edges to extend:
Select surface edges to extend:
Specify extend distance or [Modes]:
```

- The first line gives you the current values for mode and creation. AutoCAD then asks you to select the desired edge(s); when done press [Enter]. As a final step specify the extend distance.
- You can control Extension mode with the following choices:
  - Extend mode: In this mode, AutoCAD will increase the length of the surface and will try to replicate and continue the current surface shape.
  - Stretch: In this mode, AutoCAD will increase the length of the surface and will not try to replicate and continue the current surface shape.
- You can control Creation type with these choices:
  - Merge: There will be a single surface.
  - Append: There will be two surfaces, the new and the old surfaces:

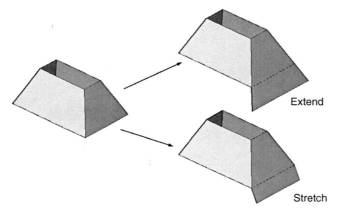

Extend

Stretch

4.4.5 Sculpt Command

- This command will help you convert a surface to a solid. To issue this command, go to the **Surface** tab, locate the **Edit** tab, and select the **Sculpt** button:

- The following prompts will be shown:

```
Mesh conversion set to: Smooth and optimized.
Select surfaces or solids to sculpt into a solid:
```

- The condition here is to have a **watertight** volume in order for the conversion process to be successful.

4.4.6 Editing Surfaces Using Grips

- You have the ability to adjust the value of the fillet radius even after the command is finished. This will work only if the curved surface created from Fillet is **not** NURBS. To do this, simply click the surface, and you will see the Fillet icon again for further editing. See the following illustration:

- You can also use grips to edit a surface created from Extrusion (which will be discussed in the next chapter). You can manipulate both the taper angle and height. See the following two illustrations:

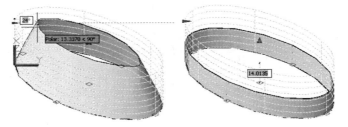

- You also have the ability to change the blend, patch, and offset values. See the following example:

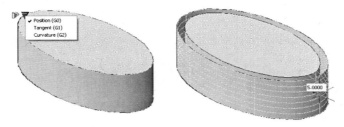

- Using Quick Properties and Properties, you can tell how the surface was created. Was it created using Extrusion or Offset? Is it Associative or NURBS? See the following examples of both the Quick Properties and Properties palette, which show the origin of the surface:

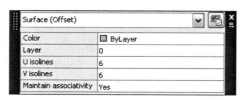

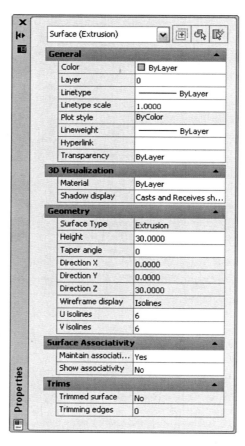

- In Properties you can change the Height, Taper angle, whether to maintain Associativity, and if this surface should be trimmed or not (this will tell you if the Untrim command will work with this surface or not).

HOW TO EDIT SURFACES

Practice 4-3

1. Start AutoCAD 2013.
2. Open **Practice 4-3.dwg**.
3. Make sure Surface Associativity is on.
4. Select the whole shape.
5. Using the arrow at the left, change the taper angle to be 20°.
6. Increase the height to 25 units.
7. Using the Extend command, extend the four lower edges (each one in a separate command) using the Stretch and Merge options by 15 units.
8. Using the Blend command, blend between the edges of the extended surfaces.
9. Thaw the 2D Shapes layer and trim using the polygon in the two directions.
10. Try to change one of the two shapes and see how the trimmed shape changes accordingly. Undo what you did.
11. Try to erase one of the two shapes and see how the trimming disappears. Undo what you did.
12. Freeze layer 2D Shapes.
13. Using the Patch command, make a cover at the top and at the bottom using the default values.
14. Try to convert the surface to a solid using the Sculpt command. The operation will not work due to the openings.
15. Untrim the openings.
16. Use Sculpt again to convert it into a solid. Did it work?
17. The final shape should look like the following:

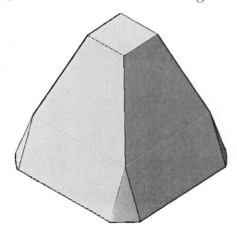

18. Save and close the file.

FILLET AND GRIPS EDITING

 Practice 4-4

1. Start AutoCAD 2013.
2. Open **Practice 4-4.dwg**.
3. Using the Patch command, create a cover from top to bottom for the elliptical shape using the default values.
4. Using the Fillet command fillet the top and bottom using radius = 5 units.
5. If a polyline is shown, don't erase it, keep it as is.
6. Using grips, change the two values to 7 units.
7. Your final shape should look like the following:

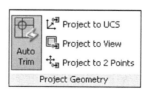

8. Save and close the file.

4.5 PROJECT GEOMETRY

- AutoCAD allows you to project 2D shapes on 3D surfaces or solids. This projected shape can then trim the surfaces or solids. You will find all of these commands in the **Surface** tab and the **Project Geometry** panel, just like the following:

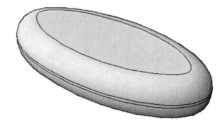

- As you can see, the big button **Auto Trim** is the on/off button. If it is on, then the projection will trim the surface or solid. The three commands are:
  - Project to UCS
  - Project to View
  - Project to 2 Points

4.5.1 Using Project to UCS

- This option projects the 2D shape along the +ve/–ve Z-axis of the current UCS:

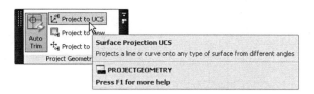

- You will see the following prompts:

```
Select curves, points to be projected or [PROjection direction]:
Select a solid, surface, or region for the target of the projection
```

- The first prompt asks you to select the desired 2D shape; you can select several objects, and when done press [Enter]. The second prompt asks you to select the desired solid or surface (you can select a 2D region as well).
- See the following illustration (notice the current UCS):

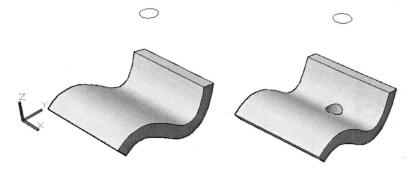

4.5.2 Using Project to View

- This option projects the geometry based on the current view:

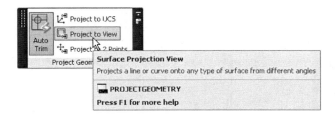

- You will see the following prompts:

```
Select curves, points to be projected or [PROjection direction]:
Select a solid, surface, or region for the target of the projection:
```

- These are identical to the first option prompts, discussed above. See the following illustration and notice how the trimming process was not affected by the current UCS:

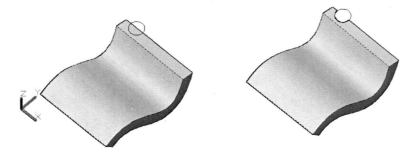

4.5.3 Using Project to 2 Points

- This option projects the geometry along a path between two points:

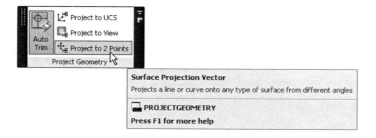

- You will see the following prompts:

```
Specify start point of direction:
Specify end point of direction:
Select curves, points to be projected or [PROjection direction]:
Select a solid, surface, or region for the target of the projection:
```

- In the first two prompts AutoCAD asks you to select the two points representing the path. The next two prompts are identical to what we saw previously.

- See the following illustration. Note that AutoCAD doesn't ask for a line to be drawn, but to illustrate the concept we draw a line here:

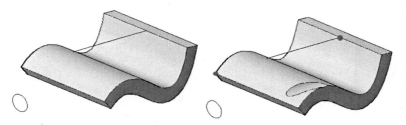

PROJECT GEOMETRY

Practice 4-5

1. Start AutoCAD 2013.
2. Open **Practice 4-5.dwg**.
3. Using the three surfaces and the three circles from left to right and the Project Geometry methods and panel, project the three circles on the three surfaces using the three methods you learned, making sure that the Auto Trim button is always on (for the third situation, the line extends from the upper-left corner to the lower-right corner).
4. Save and close the file.

4.6 CREATING NURBS SURFACES

- The most notable thing about a NURBS surface is that it has Control Vertices (CVs). These vertices give you the power to shape the surface in a very irregular way. To create a NURBS surface, go to the **Surface** tab, locate the **Create** panel, and then make sure that the **NURBS Creation** button is turned on:

- To show CVs, go to the **Surface** tab, locate the **Control Vertices** panel, and select the **Show CV** button:

- You will see something like the following:

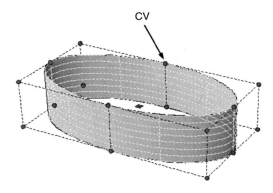

- If you click one of the CVs in order to edit it, AutoCAD will show you the following message:

- AutoCAD is telling you that in order to edit the CVs of this surface you have to convert it to Degree 3. Degree 3 means the surface can have two bends, Degree 2 means one bend, and Degree 1 means straight segments. And because the example here is a curved surface, AutoCAD wants to make sure that your surface will act right when you start editing the CVs, so it asks you to use the **Rebuild** command to change the degree to 3 in both directions.

- To issue this command, go to the **Surface** tab, locate the **Control Vertices** panel, and select the **Rebuild** button:

- The following prompt will be displayed:

```
Select a NURBS surface or curve to rebuild:
```

- Select the desired surface, and the following dialog box will be displayed:

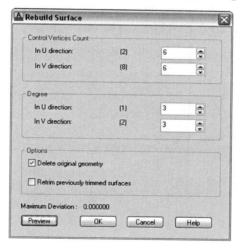

- Change the number of vertices in both directions, and change the degree value in both directions.
- These are the directions of our example before rebuilding:

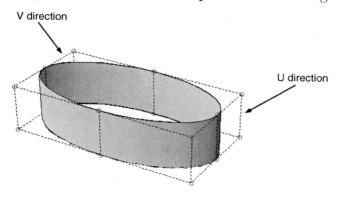

- As you can see, there are eight CVs in the V direction and two CVs in the U direction. After rebuilding, and since you change the degree to 3, AutoCAD will not accept the two vertices in the U direction, as the minimum is four vertices. See the following, which has four vertices in the U direction and eight vertices in the V direction:

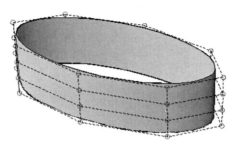

- The first technique to alter the shape is to click one of the CVs, then move it. If you hold [Shift] you can select multiple CVs. See the following, which shows how to move four vertices in one step:

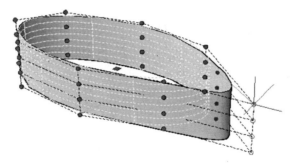

- As discussed, you can increase/decrease the number of vertices along both directions using the Rebuild command. With this command, you don't control the location of the new vertices, but using the **Add** command, you add vertices at the exact desired locations. To do that, go to the **Surface** tab, locate the **Control Vertices** panel, and select the **Add** button:

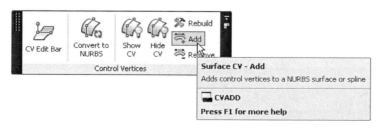

- The following prompt will be shown:

```
Select a NURBS surface or curve to add control vertices:
```

- You should select the desired location. See the following illustration:

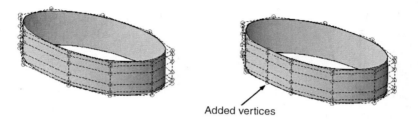

Added vertices

- You can also remove CVs. To do so, go to the **Surface** tab, locate the **Control Vertices** panel, and select the **Remove** button:

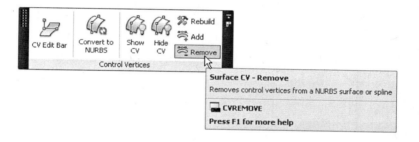

- You will see the following prompt:

```
Select a NURBS surface or curve to remove control vertices:
```

- You should select the desired set of CVs to be removed. See the following illustration:

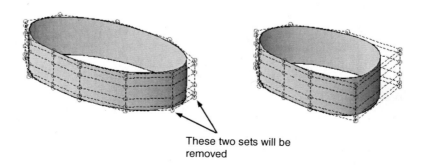

These two sets will be removed

- You can hide CVs by going to the **Surface** tab, locating the **Control Vertices** panel, and selecting the **Hide CV** button:

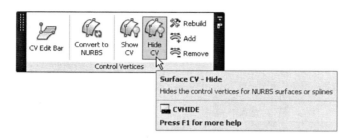

- AutoCAD allows you to convert any Procedural surface to a NURBS surface. In order to do this, go to the **Surface** tab, locate the **Control Vertices** panel, and then select the **Convert to NURBS** button:

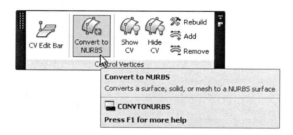

- You can add and edit CVs using the **CV Edit Bar**. To do this, go to the **Surface** tab, locate the **Control Vertices** panel, and then select the **CV Edit Bar** button:

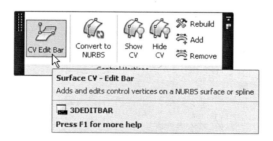

- You will see the following prompt:

```
Select a NURBS surface or curve to edit:
Select point on NURBS surface.
```

- The first prompt asks you to select the desired surface. Then specify a point on a NURBS surface to locate the new vertex location, and you will see something like the following:

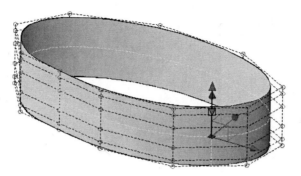

- Click the small triangle and you will see:

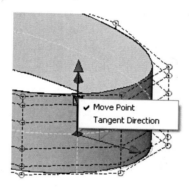

- Now you can do three things:
  - Move the new vertex:

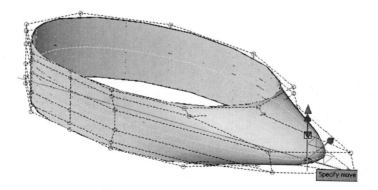

- Change the tangency:

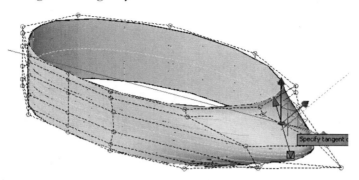

- Or use the arrow to edit the tangent magnitude:

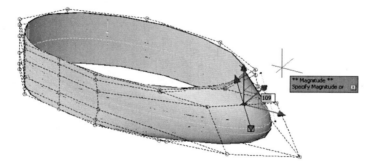

- As the final step after you finish the creation of your surfaces, you should analyze them. To this, your first step will be to turn **Hardware Acceleration** on using the status bar:

- Please note that your computer may run slower when using this option. Now go to the **Surface** tab, and you will see that there is a panel called **Analysis**, which has four buttons:

- There are three analysis methods to analyze your surface:
  - Zebra Analysis, which will test the surface continuity; see the following example. Since the continuity is G0, the two surfaces are not tangent. This is evident because the lines are not aligned on the two surfaces:

  - Curvature Analysis allows you to see Gaussian, minimum, maximum, and mean values for surface curvature. Maximum curvature and a positive Gaussian value display as green; minimum curvature and a negative Gaussian value display as blue. Planes, cylinders, and cones have zero Gaussian curvature.
  - Draft Analysis shows a color gradient onto a surface to assess whether there is sufficient space between a part and its mold.
- To control the three types of analysis, select the **Analysis Options** button, and you will see the following dialog box:

NURBS CREATION AND EDITING

Practice 4-6

1. Start AutoCAD 2013.
2. Open **Practice 4-6.dwg**.
3. This surface is a Procedural surface.
4. Convert it to a NURBS surface.
5. Show CVs.
6. Try to move one of the CVs. Does AutoCAD allow you to do so?
7. Rebuild the surface to be degree 3, with the number of vertices in U = 6 and V = 10.
8. Remove the two sets of CVs as indicated below (this is top view):

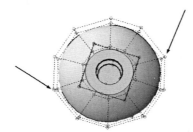

9. Go to one of the two sides you removed the CV from, and using the CV Edit bar, add a new CV almost at the center of the rectangle formed by the four CVs. Change the magnitude to be 175 moving up.
10. Do the same for the other side.
11. Patch the bottom of the shape.
12. Hide the CVs.
13. The shape will look like the following:

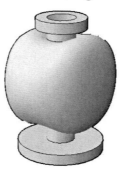

14. Save and close the file.

NOTES:

CHAPTER REVIEW

1. Using the Offset command to offset surfaces, you can offset on both sides using the same command.
    a. True
    b. False
2. In the Extend command, which one of the following is *not* an option?
    a. Extend
    b. Append
    c. Stretch
    d. Scale
3. In order to manipulate the CVs of a curved surface the degree should be at least _____.
4. Which one of the following statements is *not* true?
    a. Degree 1 means straight segments
    b. Degree 2 means two bends
    c. Degree 3 means two bends
    d. Degree 2 means one bend
5. You can add CVs, but you can't remove CVs from NURBS.
    a. True
    b. False
6. _____ will test the surface continuity.
7. In the Patch command:
    a. G0 means flat surface
    b. G1 means more curved
    c. G2 means the best curvature
    d. All of the above
8. In order to trim a surface, the 2D shape should have contact with the surface.
    a. True
    b. False

CHAPTER REVIEW ANSWERS

1. a
3. three
5. b
7. d

Chapter 5

CREATING COMPLEX SOLIDS AND SURFACES

In This Chapter

◇ 3D Poly and Helix commands
◇ Extrude command
◇ Loft command
◇ Revolve command
◇ Sweep command

5.1 INTRODUCTION

- The commands discussed in this chapter can convert 2D objects to either solids or surfaces using different techniques. These commands exist in the Solid tab and in the Surface tab. The first set of commands produce solids, and the second set produces surfaces.
- These commands are:
  - Extrude
  - Loft
  - Revolve
  - Sweep
- But before we discuss these four commands, we will first look at two commands that can help draw a 3D path:
  - 3D Poly
  - Helix

5.2 3D POLYLINE AS A PATH

- The normal Polyline command will draw 2D line/arc segments in the current XY plane, which means you can't specify the Z coordinate using it. But the 3D Polyline command allows you to specify the Z coordinate, so you can draw a 3D polyline. To issue this command, go to the **Home** tab, locate the **Draw** panel, and then select the **3D Polyline** button:

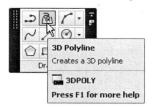

- The following prompts will be shown:

```
Specify start point of polyline:
Specify endpoint of line or [Undo]:
Specify endpoint of line or [Undo]:
Specify endpoint of line or [Close/Undo]:
```

- Compared to the normal Polyline command, this command will draw only line segments as the prompts suggest. The Polyline Edit command affects the 3D Polyline but with some differences:
  - You can't set the width using Polyline Edit.
  - You can't use the Fit option.
  - You can't use the Linetype Generation option.

5.3 HELIX AS A PATH

- This command allows you to draw a 3D spiral shape. To issue this command, go to the **Home** tab, locate the **Draw** panel, and select the **Helix** button:

- The following prompts will be shown:

```
Number of turns = 3.0000 Twist=CCW
Specify center point of base:
Specify base radius or [Diameter] <1.0000>:
Specify top radius or [Diameter] <47.1835>:
Specify helix height or [Axis endpoint/Turns/turn Height/
tWist] <1.0000>:
```

- The first line gives the current number of turns and the twist direction. AutoCAD then asks you to specify the center point of the base, radius, or diameter of the base and the top radius or diameter. Finally, you will specify the total height of the helix, and you will get something similar to the following:

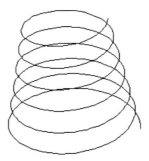

- The above are the default steps, but you can alter some of the other options in this command, which are discussed as follows.

5.3.1 Axis Endpoint Option

- By default, the base of the helix will be in the current XY plane. This option allows you to specify the base in a different plane. The following prompts will be shown:

```
Specify axis end point:
```

5.3.2 Turns Option

- This option allows you to input the number of turns for the current helix, rather than accepting the default value, and accordingly, AutoCAD will calculate the turn height after knowing the total height of the helix. The following prompts will be shown:

```
Enter number of turns:
Specify helix height or [Axis endpoint/Turns/turn Height/
tWist]:
```

5.3.3 Turn Height Option

- This option allows you to specify the turn height, and as above, AutoCAD will calculate the number of turns after knowing the total height. The following prompts will be shown:

```
Specify distance between turns:
Specify helix height or [Axis endpoint/Turns/turn Height/
tWist]:
```

5.3.4 Editing a Helix Using Grips

- After you create the helix, you can edit it using grips. If you click the helix you will see the following:

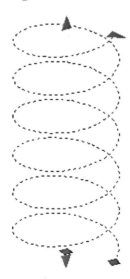

- You will see two triangles at the top pointing upwards and at the bottom pointing downwards; these allow you to increase or decrease the total height of the selected helix. The triangle at the right will increase or decrease the top radius without changing the value of the base radius. Finally, the grip at the bottom will increase or decrease the radius for both the top and bottom.

- You can also edit the helix using the Quick Properties and Properties palette:

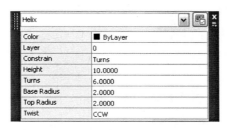

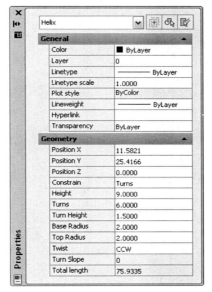

- You can edit all or any of the values of the helix.

DRAWING A 3D POLYLINE AND HELIX

Practice 5-1

1. Start AutoCAD 2013.
2. Open **Practice 5-1.dwg**.
3. Make 3D Poly the current layer.

4. Using the 3D Poly command, track the line as indicated in the following illustration using OSNAP (turn on Show/Hide Lineweight from the status bar):

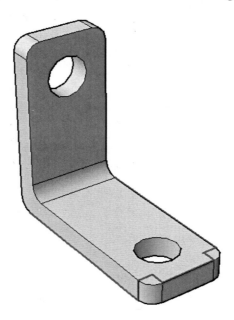

5. Make layer Helix current.
6. Draw a helix at the lower groove using the following information:
   a. Center = Center of the groove
   b. Radius of bottom and top = 7.5 (you can select using OSNAP)
   c. Leave number of turns = 3 and input height = 7.5
7. Do the same thing for the top groove (use the Axis endpoint option).
8. Save and close the file.

5.4 USING THE EXTRUDE COMMAND

- This command can produce a solid or a surface from a 2D profile perpendicular to the original 2D shape. You can use the edges of the faces of both solids and surfaces to extrude as well. Using the Extrude command in the Solid tab will produce solids, but the 2D shape should be closed in order for this process to be successful. Using the Extrude command in the Surface tab will produce surfaces, and this command will work with both open and closed 2D shapes.

■ To issue this command, go to the **Solid** tab, locate the **Solid** panel, and select the **Extrude** button, or go to the **Surface** tab, locate the **Create** panel, and select the **Extrude** button:

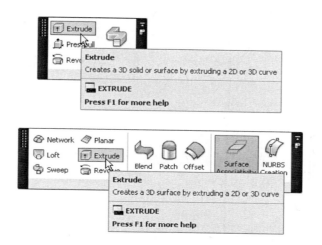

■ The following prompts will be shown:

```
Select objects to extrude or [MOde]:
Specify height of extrusion or [Direction/Path/Taper angle]
```

■ The first line asks you to select the object(s) to extrude. The second line says:

```
Specify height of extrusion or [Direction/Path/Taper angle]
```

■ This line tells you there are four ways to create an extrusion in AutoCAD:
  • Specify a height
  • Specify a direction
  • Specify a path
  • Specify a taper angle

5.4.1 Creating an Extrusion Using Height

■ This is the default option, and it will create an extrusion perpendicular to the 2D profile using height. Whenever you select the 2D object, edge, or face, you will see the extrusion on the screen, and you can specify the height graphically or you can input the height value. See the following:

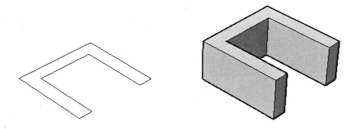

5.4.2 Creating an Extrusion Using Direction

- If the Height option is used to create an extrusion perpendicular on the profile, this option will give you the ability to create an extrusion with any angle. The following prompts will be shown:

```
Specify start point of direction:
Specify end point of direction:
```

- The following illustration clarifies the concept:

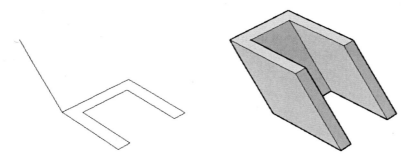

5.4.3 Creating an Extrusion Using Path

- This option will create an extrusion using a path. You can use almost all 2D objects as paths including a 3D poly and helix as well as the edges of surfaces and solids. Before selecting the path, you have the choice to set up a Taper angle. The following prompts will be shown:

```
Select extrusion path or [Taper angle]:
```

- If you select the Taper angle option, you will see the following prompts:

```
Specify angle of taper for extrusion or [Expression]:
```

- The following helps explain selecting a path:

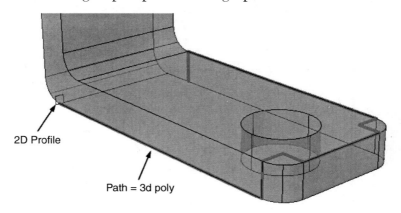

2D Profile

Path = 3d poly

- The result would look like the following:

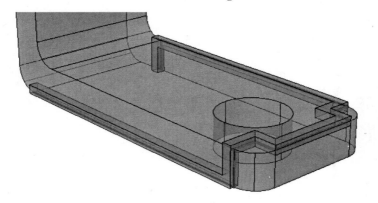

- The following example explains using path with Taper angle:

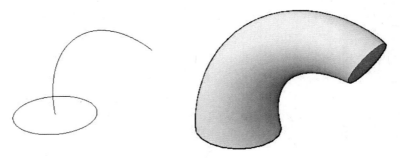

Taper angle = 5

5.4.4 Creating an Extrusion Using Taper Angle

- This option allows you to set a positive taper angle (to the inside) or a negative taper angle (to the outside). After you set the Taper angle, the same prompts will come up again; that is, Taper angle can work with all the above options (height, direction, and path). See the following prompts:

```
Specify angle of taper for extrusion or [Expression]:
Specify height of extrusion or [Direction/Path/Taper angle/
Expression]
```

Taper angle = 15

USING THE EXTRUDE COMMAND

Practice 5-2

1. Start AutoCAD 2013.
2. Open **Practice 5-2.dwg**.
3. Change the visual style to X-Ray.
4. Thaw layer 2D Object.
5. Zoom to the lower-left corner of the part, and you will notice a rectangle. Using Extrude with Path, extrude the shape using the path.
6. Subtract the newly created solid from the big shape.
7. Freeze layer 3D Poly.
8. Using the cylindrical hole at the lower part, change the subobject filter to edge, using the Extrude command, select the upper edge of the hole, and using Taper angle = −20 and height downwards by 5 units, create a solid.
9. Subtract the newly created solid from the big shape.
10. Change the visual style to Conceptual.

11. You will get the following shape:

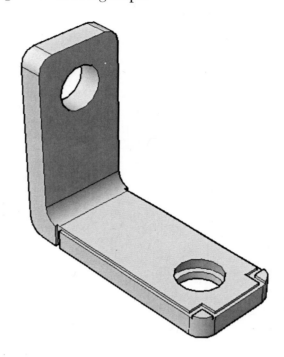

12. Save and close the file.

5.5 USING THE LOFT COMMAND

- This command allows you to create a solid or surface by using multiple cross sections in different UCSs, whether opened or closed. The condition here is to have all profiles opened or all of them closed, since you can't have a mixed selection. You can use different options such as guides, path, and cross sections. To issue this command, go to the **Solid** tab, locate the **Solid** panel, and select the **Loft** button, or go to the **Surface** tab, locate the **Create** panel, and select the **Loft** button:

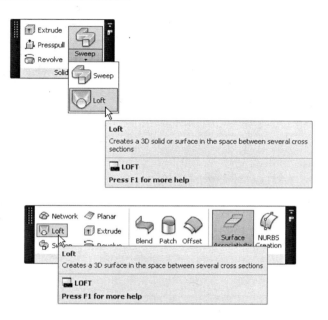

- The following prompts will be displayed:

```
Select cross sections in lofting order or [POint/Join multiple
edges/MOde]:
```

- The first line asks you to select the cross sections in lofting order. See the following example:

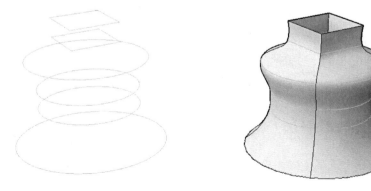

- This prompt will be repeated and allows you to select more closed or opened objects. While you are selecting you can choose one of three more options:
  - Point, which allows you to select the loft end point and end the command at this point. The cross section selected should always be closed, and you should specify this point either by inputting a real 3D point, or by selecting a point object. You will see the following prompt:

```
Specify loft end point:
```

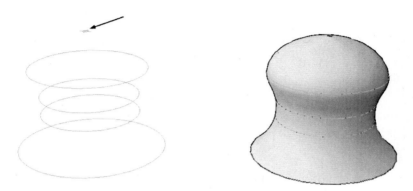

- The Join multiple edges option will join the edges of solids or surfaces to form a cross section. The following prompt will be shown:

```
Select edges that are to be joined into a single cross section:
```

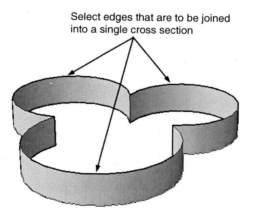

Select edges that are to be joined
into a single cross section

- The Mode option is used to select the type of resulting object, a solid or a surface.

5.5.1 Lofting Using Guides

- This type of lofting will be dictated by an object called a guide, which will determine the final shape of the solid or surface. You can select more than one guide. The following prompts will be shown:

```
Select guide curves:
```

- You will get something like the following:

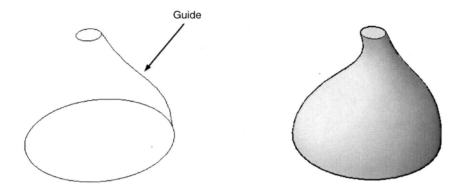

- Guide object(s) should touch all cross sections in order for the process to succeed.

5.5.2 Lofting Using Path

- You have the ability to select only cross sections, but with this option, you can select both cross sections and a path to dictate the final shape. The following prompts will be shown:

```
Select path profile:
```

- You should get something like the following:

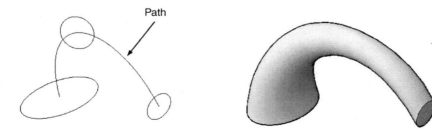

5.5.3 Lofting Using Cross Sections Only

- This option allows you to use only the selected cross sections to build the solid or surface without any additional objects. Once you select this option the command will end, and you should select the resulting solid or surface to show a small triangle at one of the edges of the shape. When you click it, it will show different options like the following:

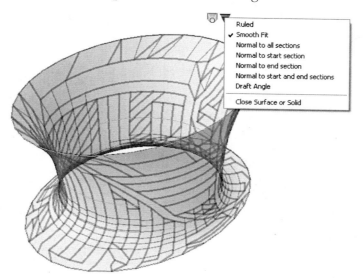

- The default option is Smooth Fit. The Ruled option allows you to produce a solid or surface by connecting the cross sections without smoothing. The following illustration compares the two results:

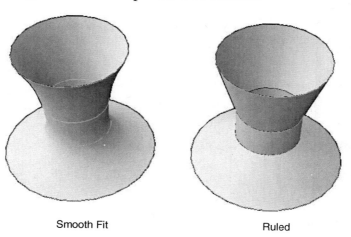

Smooth Fit   Ruled

- The other four options all start with the word Normal. Normal means the curve is 90° to the cross section, but what does the cross section mean? Here is the answer:
  - Normal to all sections
  - Normal to start section
  - Normal to end section
  - Normal to both start and end sections
- See the following example:

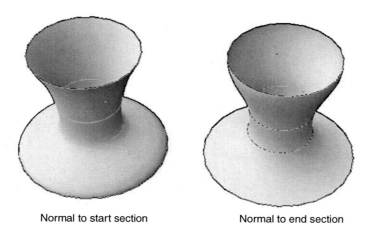

Normal to start section   Normal to end section

- The last option is Draft Angle, which allows you to change the angle of the curve related to the first and last cross sections. See the following illustration:

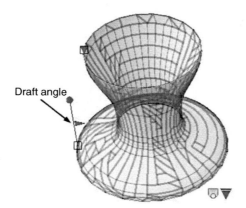

USING THE LOFT COMMAND

Practice 5-3

1. Start AutoCAD 2013.
2. Open **Practice 5-3.dwg**.
3. Using the shape at the left, loft the three circles using the path.
4. Using the six circles at the right, loft them using the cross sections only. As you can see the default is Smooth Fit; check the Ruled options.
5. You can make copies of the six circles and see the difference between all the other Normal options.
6. Check all Normal options.
7. Check the two draft angles.
8. Finalize the shape using Normal to all sections.
9. You should have something like the following:

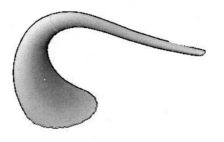

10. Save and close the file.

5.6 USING THE REVOLVE COMMAND

- This command allows you to create a solid or surface by revolving 2D objects, closed or opened, or using the face or edge of a solid or surface. To issue this command, go to the **Solid** tab, locate the **Solid** panel, and select the

Revolve button, or go to the **Surface** tab, locate the **Create** panel, and select the **Revolve** button:

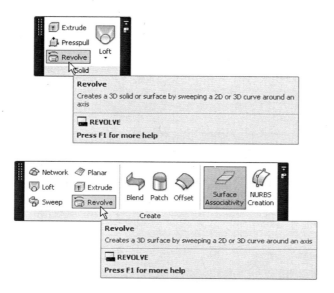

- The following prompts will be shown:

```
Select objects to revolve or [MOde]:
```

- AutoCAD asks you to select the desired object(s) to revolve. When done press [Enter], and the following prompts will be shown:

```
Specify axis start point or define axis by [Object/X/Y/Z]
<Object>:
Specify angle of revolution or [STart angle/Reverse/EXpression]
<360>:
```

- AutoCAD offers three different methods to specify the axis of revolution:
  - Specify an axis by inputting two points.
  - Specify an axis by selecting a drawn object.
  - Specify an axis by specifying one of the three axes.
- Finally, you have to specify the angle of revolution; there are multiple options here:
  - Input the angle either by typing or using the mouse. CCW is always positive. If you define the axis of revolution by specifying two points, then the positive angle will be the curling of your right hand, with your thumb pointing to the direction of the nearest point to the farthest point.

- Specify the start angle if you don't want the revolving process to start from the cross section.
- If you want to reverse the direction of the angle, use the Reverse option.
■ The following illustration helps explain the concept:

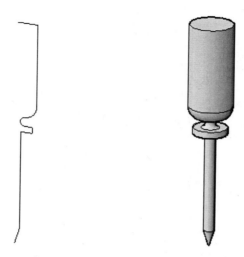

USING THE REVOLVE COMMAND

Practice 5-4

1. Start AutoCAD 2013.
2. Open **Practice 5-4.dwg**.
3. Revolve the shape at the left using two points as the axis of revolution, choosing from top to bottom and assigning 180° as the rotation angle.
4. Undo the last step.
5. Revolve the shape at the left using two points as the axis of revolution, choosing from bottom to top and assigning 180° as the rotation angle. Did you notice the difference? Why? _____
6. Revolve the shape at the right using the line as the axis of revolution using 360° as the rotation angle.

7. The final result should look like the following:

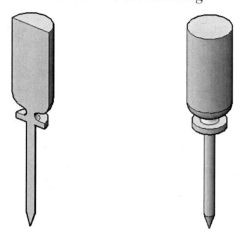

8. Save and close the file.

5.7 USING THE SWEEP COMMAND

- This command allows you to create a complex solid or surface by sweeping a 2D, open or closed shape or a solid or surface face or edge along a path. To issue this command, go to the **Solid** tab, locate the **Solid** panel, and select the **Sweep** button, or go to the **Surface** tab, locate the **Create** panel, and select the **Sweep** button:

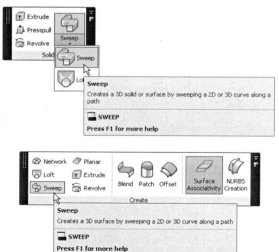

- The following prompts will be shown:

```
Select objects to sweep or [MOde]:
Select objects to sweep or [MOde]:
Select sweep path or [Alignment/Base point/Scale/Twist]:
```

- The default option is to select the object(s) to sweep and then select the sweep path if you follow this path; the following is the result:

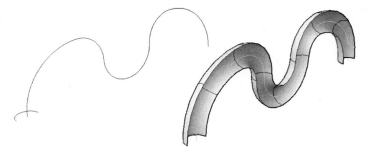

- Or you can use the other options:
  - Alignment
  - Base point
  - Scale
  - Twist

5.7.1 Alignment Option

- This option sets the relationships between the profile and the path. Sometimes the path is normal to the profile, and sometimes it is not! See the following two examples. In the first one the path is normal to the profile:

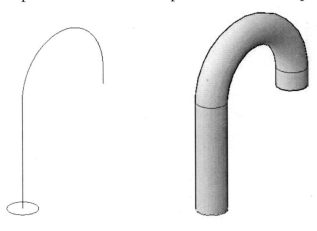

- And in the other example, the path is not normal to the profile:

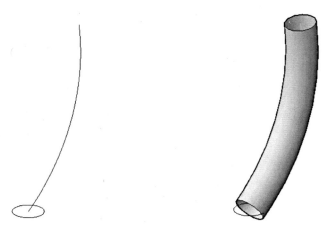

- In the second example, if you turn Alignment off, you will get the following result:

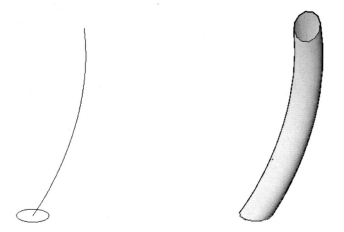

- You will see the following prompt:

```
Align sweep object perpendicular to path before sweep [Yes/
No]<Yes>:
```

5.7.2 Base Point Option

- This option allows you to specify a new base point for the solid or surface produced other than the location of the profile object. You will see the following prompt:

```
Specify base point:
```

- Either input the coordinates, or specify them using the mouse. The following example explains the concept:

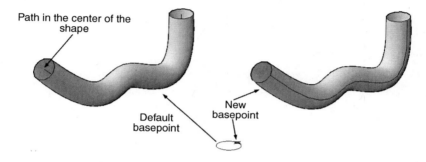

Path in the center of the shape

Default basepoint

New basepoint

5.7.3 Scale Option

- This option allows you to set a scaling factor for the end of the sweeping solid or surface. You will see the following prompt:

```
Enter scale factor or [Reference/Expression]<1.0000>:
```

- Input the desired scale, then press [Enter], and the result will look something like the following:

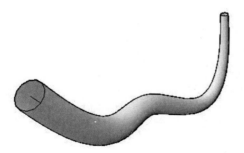

5.7.4 Twist Option

- This option allows you to set the angle of the twist. This command will perform while sweeping. You will see the following prompt:

```
Enter twist angle or allow banking for a non-planar sweep
path [Bank/EXpression]<0.0000>:
```

- Input the desired twist angle, and the result will look something like the following:

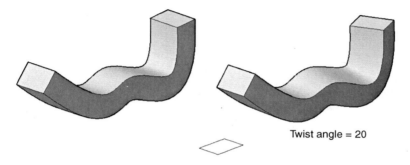

Twist angle = 20

USING THE SWEEP COMMAND

 Practice 5-5

1. Start AutoCAD 2013.
2. Open **Practice 5-5.dwg**.
3. Thaw layer Helix and layer 2D Object. Turn on Show/Hide Lineweight.
4. Zoom in to the groove at the bottom.
5. Using the Sweep command, use the circle as the profile and the helix as the path without changing the defaults.
6. Subtract the new shape from the solid.
7. Repeat the same for the top groove, but this time use scale = 0.5.
8. Freeze layer Helix.
9. Zoom in to the lower-left corner of the top shape to see the rectangle there.
10. Using the Sweep command (from the Surface tab), use the rectangle as the profile and the edge of the solid as the path (you will need to change the Subobjects filter to Edge).
11. Repeat the same thing on the other side.

12. The final shape should look like the following:

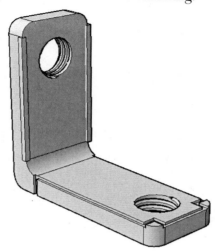

13. Save and close the file.

5.8 EDITING SOLIDS AND SURFACES CREATED USING THE FOUR COMMANDS

- After creating solids or surfaces, clicking the objects allows you to edit them in three ways:
  - Using Grips
  - Using Quick Properties
  - Using Properties

5.8.1 Using Grips

- If you clicked a solid or a surface created from Extrusion, you will see the following:

To change the taper angle

To change the height

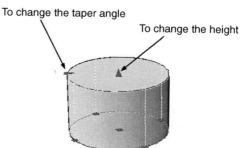

- If you click a solid or a surface created from Loft, you will see the following:

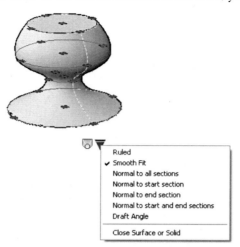

- If you click a solid or a surface created from Revolve, you will see the following:

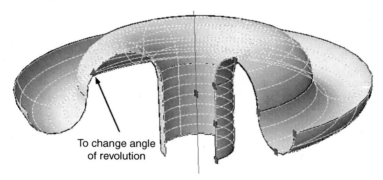

To change angle of revolution

- If you click a solid or surface created from Sweep, you will see the following:

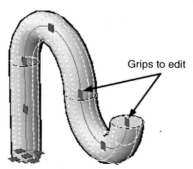

Grips to edit

5.8.2 Using Quick Properties

- Quick Properties shows different types of information for solids and surfaces. The following two examples are for surfaces:

- On the other hand, you will see the following for solids:

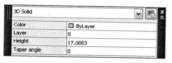

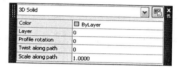

5.8.3 Using Properties

- As for Properties palette, you will see the following:

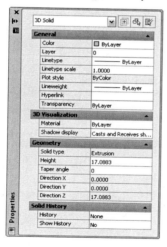

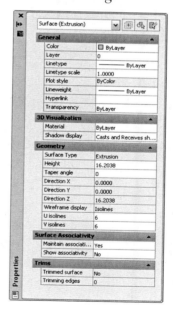

- It's easy to see that Quick Properties and Properties reveal more information for surfaces than for solids.

5.8.4 DELOBJ System Variable

- Using 2D objects in Extrude, Loft, Revolve, and Sweep will produce either a solid or surface object, but what will happen to this 2D object? Will it stay? Will it be erased? To answer this we have to introduce the system variable DELOBJ, which controls whether to delete the defining 2D object. If DELOBJ = 0, all of the defining 2D objects such as profiles and cross sections, paths, and guide curves will stay. If DELOBJ = 1, only profiles will be deleted. If DELOBJ = 2, then profiles, paths, and guide curves will be deleted.

PROJECT USING METRIC UNITS

Project 5-1-M

1. Start AutoCAD 2013.
2. Open **Project 5-1-M.dwg**.
3. For each part of the drawing there is a layer with the same name; make sure you are in the right layer when you draw the shape.
4. Draw a box 30x30x0.9m.
5. At the center of the top face of the box draw a cylinder with R = 3m and height = 6m.
6. Using one of the quadrants of the top face of the cylinder, draw a rectangle 0.15x0.15. If it was drawn outside, move it to the inside using the midpoint of one of the sides to coincide with the quadrant of the cylinder.
7. Using the Surface tab, extrude the rectangle by 0.6m.
8. Using Polar Array, create an array from the small columns with the angle between each shape as 30°.
9. You will have the following shape:

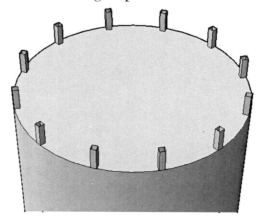

10. Draw a line between the top midpoints of any two columns.
11. Using the gizmo move it down by 0.3m.
12. Draw the line again.
13. At the end of one of the lines, draw a circle with R = 0.01m. Do the same to the other line.
14. Using the Surface tab, sweep the circle using the line as your path. Repeat the same for the other line.
15. Using Polar Array and the same input you used for the columns, array the protecting ropes.
16. The shape should look like the following:

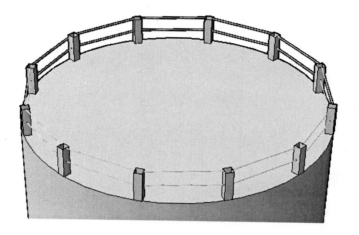

17. Draw a circle at the center of the top face of the cylinder with R = 0.6m.
18. Using Press/Pull, pull it upwards by 0.12m.
19. At the top of the new shape, draw another circle with R = 0.3m.
20. Change the UCS to look like the following:

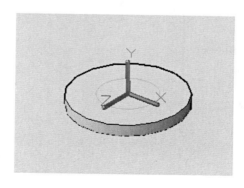

21. Draw the following polyline:

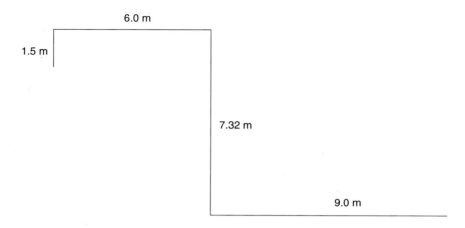

6.0 m

1.5 m

7.32 m

9.0 m

22. Using the Fillet command, fillet the polyline using radius = 0.5 m and the Polyline option to do it in one step.
23. Using the Surface tab, extrude the circle with R = 0.3 using the polyline as your path.
24. The shape should look like the following:

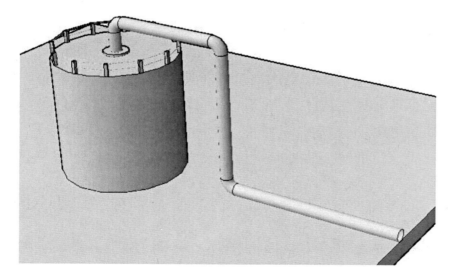

25. Draw a helix with its center point at the center point of the cylinder, with radius at the bottom and at the top = 3.2 m and height = 6 m.

26. Locate the start of the helix at the top, and at that position erase the two protecting ropes.
27. Draw a box with height = 0.02 at its two opposite corners at two corners of the two columns, as follows:

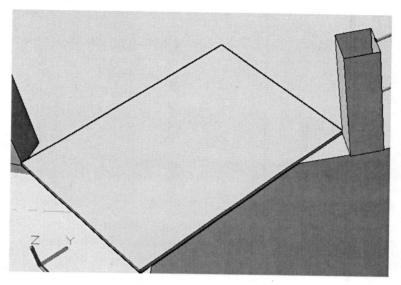

28. Using the outer lower edge (you will need to change the Subobject filter to Edge) of the box as the profile and the helix as your path, sweep the shape using the Surface tab.
29. The final shape should look like the following:

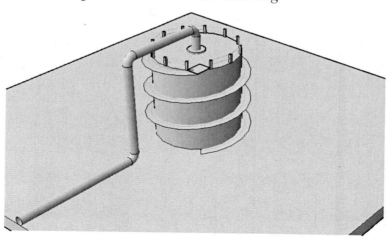

30. Save and close the file.

PROJECT USING IMPERIAL UNITS

Project 5-1-I

1. Start AutoCAD 2013.
2. Open **Project 5-1-I.dwg**.
3. For each part of the drawing there is a layer with the same name; make sure you are in the right layer when you draw the shape.
4. Draw a box 100'x100'x3'm.
5. At the center of the top face of the box draw a cylinder with R = 10' and height = 20'.
6. Using one of the quadrants of the top face of the cylinder, draw a rectangle 6"x6". If it was drawn outside, move it to the inside using the midpoint of one of the sides to coincide with the quadrant of the cylinder.
7. Using the Surface tab, extrude the rectangle by 2'.
8. Using Polar Array, create an array from the small columns with the angle between each shape as 30°.
9. You will have the following shape:

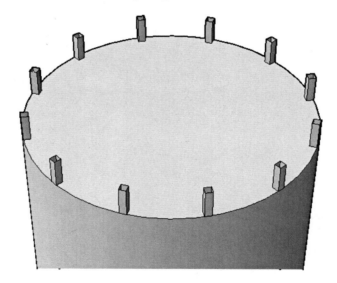

10. Draw a line between the top midpoints of any two columns.
11. Using the gizmo, move it to down by 1'.
12. Draw the line again.
13. At the end of one of the lines, draw a circle with R = 0.4". Do the same to the other line.

14. Using the Surface tab, sweep the circle using the line as your path. Repeat the same for the other line.
15. Using Polar Array and the same input you used for the columns, array the protecting ropes.
16. The shape should look like the following:

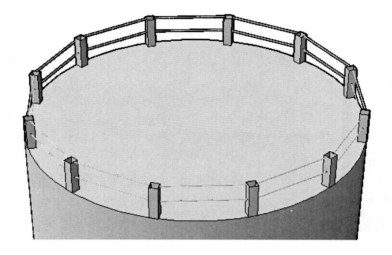

17. Draw a circle at the center of the top face of the cylinder with R = 2'.
18. Using Press/Pull, pull it upwards by 5".
19. At the top of the new shape, draw another circle with R = 1'.
20. Change the UCS to look like the following:

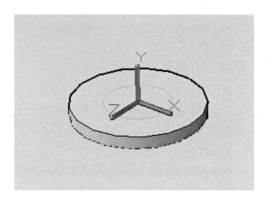

21. Draw the following polyline:

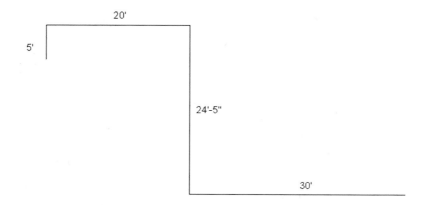

22. Using the Fillet command, fillet the polyline using radius = 20" and the Polyline option to do it in one step.
23. Using the Surface tab, extrude the circle with R = 1' using the polyline as your path.
24. The shape should look like the following:

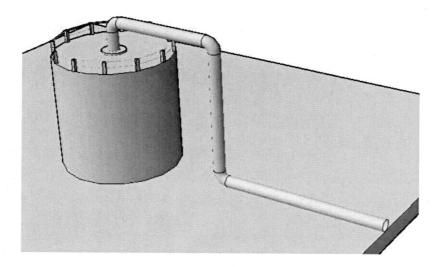

25. Draw a helix with its center point at the center point of the cylinder and with the radius at the bottom and the top = 10'-8" and height = 20'.
26. Locate the start of the helix at the top, and at that position erase the two protecting ropes.

27. Draw a box with height = 0.8" at its two opposite corners at two corners of the two columns, as follows:

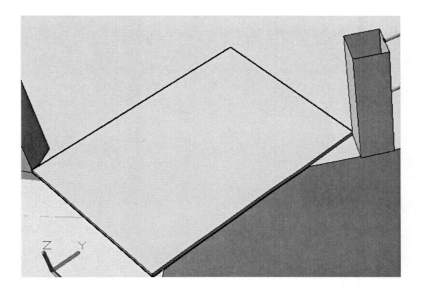

28. Using the outer lower edge (you will need to change the Subobject filter to Edge) of the box as the profile and the helix as your path, sweep the shape using the Surface tab.
29. The final shape should look like the following:

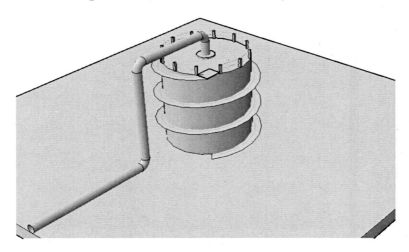

30. Save and close the file.

NOTES:

CHAPTER REVIEW

1. Which one of the following statements is false?
   a. DELOBJ has three different values.
   b. Quick Properties and Properties will show different information for surfaces and solids.
   c. You can't consider a line a sweeping path.
   d. You can delete the profile after finishing the revolution using the DELOBJ system variable.
2. Twist and scale are options in the _____ command.
3. All of the following are options in the Loft command except which one?
   a. Path
   b. Guide
   c. Normal to all sections
   d. Normal to no section
4. While using the Revolve command, the axis of revolution should always be a drawn object.
   a. True
   b. False
5. While you are using the grips of an extrusion you can change the taper angle.
   a. True
   b. False
6. The default twist of a helix is _____.

CHAPTER REVIEW ANSWERS

1. c
3. d
5. a

6

Chapter

SOLID EDITING COMMANDS

In This Chapter

◇ Solid face manipulation commands
◇ Solid edge manipulation commands
◇ Solid body manipulation commands

6.1 INTRODUCTION

- In AutoCAD there is a specific command that can be used to edit everything related to a solid object: edges, faces, and the whole body of the solid. This command is called **solid edit**, and you access it using the Command window. AutoCAD also included all the inner options in the Ribbon interface. If you go to the **Home** tab and locate the **Solid Editing** panel, you will see the face commands:

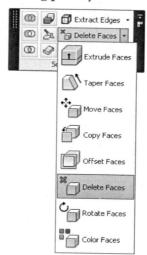

- Edge commands:

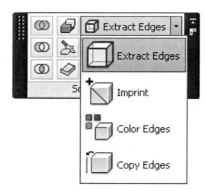

- Body commands:

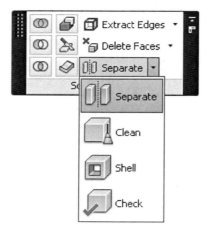

- In the following pages we will discuss all of these commands.

6.2 EXTRUDING SOLID FACES

- This command allows you to extrude selected *planar* faces of a solid. You can use different methods for extruding a face such as using height, path, and taper angle. If you input a positive value AutoCAD will increase the volume of the solid, and a negative value will decrease the volume of the solid (this works backwards if you want to increase an opening in a solid).

- To issue this command, go to the **Home** tab, locate the **Solid Editing** panel, and then select the **Extrude Faces** button:

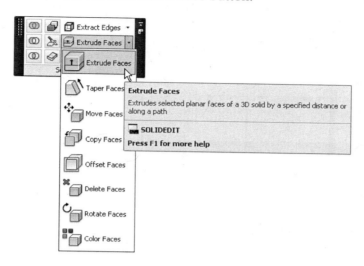

- You will see the following prompts:

```
Select faces or [Undo/Remove]:
Select faces or [Undo/Remove/ALL]:
Specify height of extrusion or [Path]:
Specify angle of taper for extrusion <0>:
Solid validation started.
Solid validation completed.
```

- The best way to select solid face(s) is to hold [Ctrl], then select the desired face(s) or change the subobjects to Face. The rest of the prompts were already discussed with the Extrude command. You have to press [Enter] twice to end this command. The following helps explain the concept:

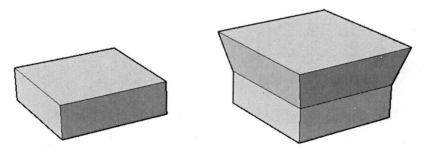

6.3 TAPERING SOLID FACES

- This command allows you to create an inclined face using two points. A positive angle means the face will incline inwards, and a negative angle will incline outwards. To issue this command, go to the **Home** tab, locate the **Solid Editing** panel, and then select the **Taper Faces** button:

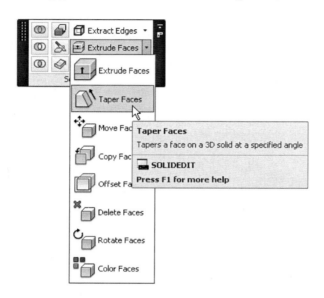

- The following prompts will be shown:

```
Select faces or [Undo/Remove]:
Select faces or [Undo/Remove/ALL]:
Specify the base point:
Specify another point along the axis of tapering:
Specify the taper angle:
Solid validation started.
Solid validation completed.
```

- Make sure Faces is selected at the filter pull-down in the Subobject panel to make the selection process easier, or hold the [Ctrl] key before selecting the desired face(s). After selecting the desired faces, specify a base point and a second point to define the axis of tapering, then input the tapering angle, then press [Enter] twice. See the following example (we used a positive angle, so the face went to the inside):

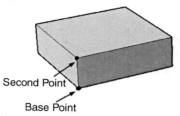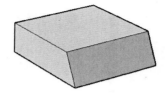

Second Point

Base Point

EXTRUDING AND TAPERING SOLID FACES

Practice 6-1

1. Start AutoCAD 2013.
2. Open **Practice 6-1.dwg**.
3. Extrude the face of the small cylinder using the path.
4. Using the Taper Face command, taper the face of the small cylinder using the quadrant at the left and making sure Polar is on to get a straight line (the angle is 12°).
5. The final result should look like the following:

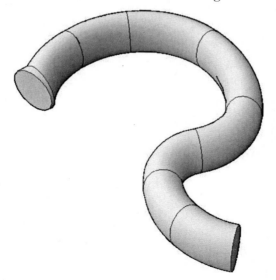

6. Save and close the file.

6.4 MOVING SOLID FACES

- This command will help you move the face(s) of a solid using a base point. To issue this command, go to the **Home** tab, locate the **Solid Editing** panel, and then select the **Move Faces** button:

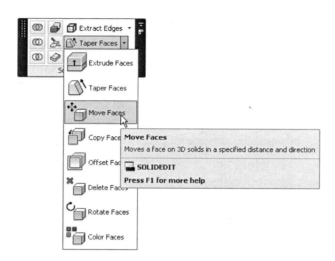

- You will see the following prompts:

```
Select faces or [Undo/Remove]:
Select faces or [Undo/Remove/ALL]:
Specify a base point or displacement:
Specify a second point of displacement:
Solid validation started.
Solid validation completed.
```

- Make sure you are using the Face option in the Subobject panel to select the face(s) or hold [Ctrl] before selecting the desired face(s), then specify a base point and a second point (identical to moving in 2D). Finally, press [Enter] twice to end the command.

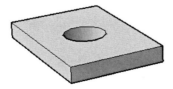 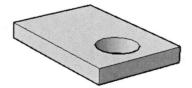

6.5 COPYING SOLID FACES

- This command will help you copy the face(s) of a solid using a base point to produce a region. To issue this command, go to the **Home** tab, locate the **Solid Editing** panel, and then select the **Copy Faces** button:

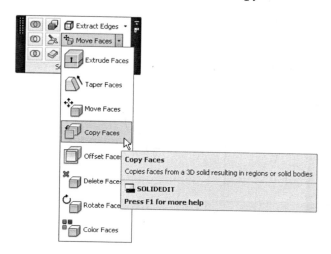

- You will see the following prompts:

```
Select faces or [Undo/Remove/ALL]:
Select faces or [Undo/Remove/ALL]:
Specify a base point or displacement:
Specify a second point of displacement:
Enter a face editing option
```

- Make sure you are using the Face option in the Subobject panel to select face(s) or hold [Ctrl] before selecting the desired face(s), then specify a base point and a second point (identical to copying in 2D). Finally, press [Enter] twice to end the command. The result will be a region object for planar faces and surfaces for curved faces:

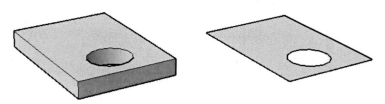

MOVING AND COPYING SOLID FACES

 Practice 6-2

1. Start AutoCAD 2013.
2. Open **Practice 6-2.dwg**.
3. Move the hole from its center to the center of the curved shape.
4. Copy the following two faces:

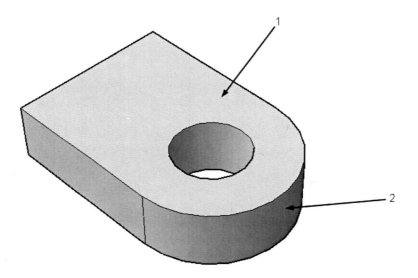

5. What is the object type of the copied face from (1)? _____ (Region)
6. What is the object type of the copied face from (2)? _____ (Surface)
7. Save and close the file.

6.6 OFFSETTING SOLID FACES

- This command allows you to offset solid faces. The Offset command is different than the Extrude command because the Offset command can offset both planar and curved faces. If you input a positive offset distance, the volume will increase, but if you input a negative offset distance the volume will decrease. To issue this command, go to the **Home** tab, locate the **Solid Editing** panel, and then select the **Offset Faces** button:

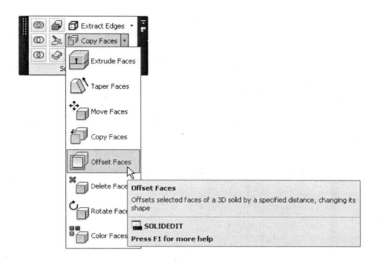

- The following prompts will be shown:

```
Select faces or [Undo/Remove]:
Select faces or [Undo/Remove/ALL]:
Specify the offset distance:
Solid validation started.
Solid validation completed.
```

- Make sure you are using the Face option in the Subobject panel to select face(s) or hold [Ctrl] before selecting the desired face(s), then specify the offset distance. Finally, press [Enter] twice to end the command.

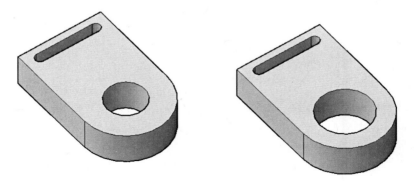

6.7 DELETING SOLID FACES

- This command allows you to delete a solid face, but this process does not always work. If you delete a face, AutoCAD will not be able to close its gap, so the process will not continue, and you will see the following prompt:

```
Modeling Operation Error:
Gap cannot be filled.
```

- To issue this command, go to the **Home** tab, locate the **Solid Editing** panel, and then select the **Delete Faces** button:

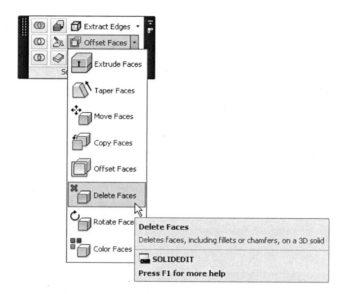

- The following prompts will be shown:

```
Select faces or [Undo/Remove]:
Select faces or [Undo/Remove/ALL]:
Solid validation started.
Solid validation completed.
```

- Make sure you are using the Face option in the Subobject panel to select the face(s) or hold [Ctrl] before selecting the desired face(s), and then select the desired faces to be deleted. Finally, press [Enter] twice to end the command.

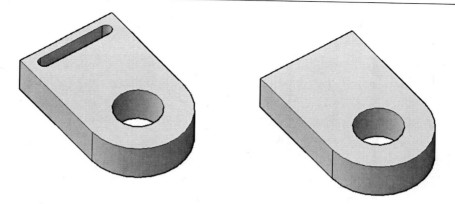

OFFSETTING AND DELETING SOLID FACES

Practice 6-3

1. Start AutoCAD 2013.
2. Open **Practice 6-3.dwg**.
3. Offset the cylindrical hole to be larger by 5 units.
4. Offset the following two faces to the outside by 5 units:

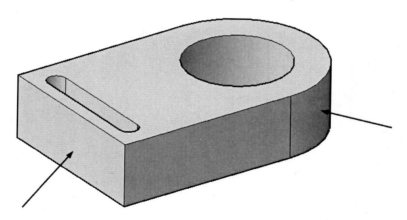

5. Delete the small openings.
6. The final shape should look like the following:

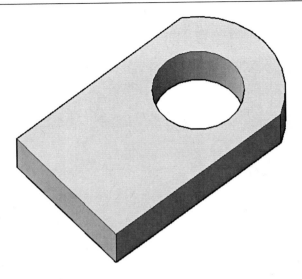

7.   Save and close the file.

6.8 ROTATING SOLID FACES

- This command allows you to rotate a solid face. To issue this command, go to the **Home** tab, locate the **Solid Editing** panel, and select the **Rotate Faces** button:

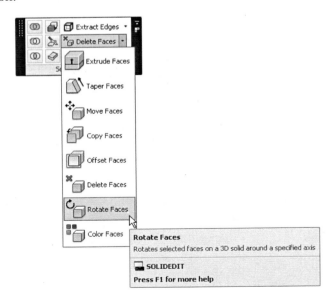

- The following prompts will be shown:

```
Select faces or [Undo/Remove]:
Select faces or [Undo/Remove/ALL]:
Specify an axis point or [Axis by object/View/Xaxis/Yaxis/
Zaxis] <2points>:
Specify the second point on the rotation axis:
Specify a rotation angle or [Reference]:
Solid validation started.
Solid validation completed.
```

- Make sure you are using the Face option in the Subobject panel to select the face(s) or hold [Ctrl] before selecting the desired face(s), then select the desired faces to be rotated, and specify the axis of rotation using any of the following methods:
  - Axis by Object
  - View
  - X-axis, Y-axis, Z-axis
  - 2 points
- The last step is to specify the rotation angle. The positive angle is in the direction of the curling fingers of your right hand, while your thumb is pointing to the positive part of the axis of rotation. If you specified two points, then from point 1 to point 2 is the positive part of the axis of rotation. Finally, press [Enter] twice to end the command.
- See the illustration below:

6.9 COLORING SOLID FACES

- This command allows you to assign the face(s) of a solid a color. To issue this command, go to the **Home** tab, locate the **Solid Editing** panel, and then select the **Color Faces** button:

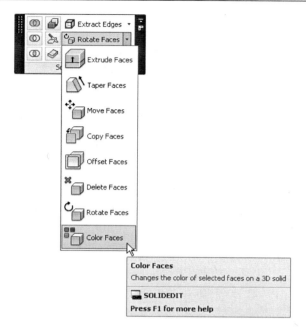

- The following prompts will be shown:

```
Select faces or [Undo/Remove]:
Select faces or [Undo/Remove/ALL]:
```

- Make sure you are using the Face option in the Subobject panel to select the face(s) or hold [Ctrl] before selecting the desired face(s), and then select the desired faces to be colored and press [Enter]. The Select Color dialog box will open; select your desired color and click OK.

ROTATING AND COLORING SOLID FACES

Practice 6-4

1. Start AutoCAD 2013.
2. Open **Practice 6-4.dwg**.
3. Using Press/Pull, pull the right part of the top face of the cylinder upwards by 6 units.
4. Rotate the two faces by 15° to produce the following shape:

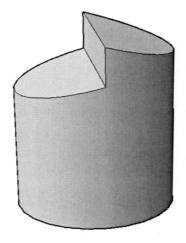

5. Color the two faces using the color red.
6. Save and close the file.

6.10 EXTRACTING SOLID EDGES

- This command allows you to create a wireframe of lines from the edges of a selected solid. To issue this command, go to the **Home** tab, locate the **Solid Editing** panel, and then select the **Extract Edges** button:

- The following prompts will be shown:

```
Select objects:
Select objects:
```

- AutoCAD asks you to select the desired solid(s). When done press [Enter]. The output will be at the same place of the original solid, so you should use the Move command to move the solid, so the extracted edges will be displayed. The following helps illustrate the concept:

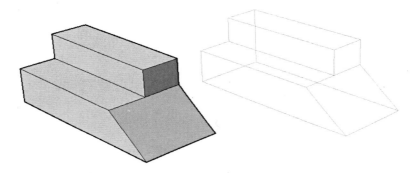

6.11 EXTRACT ISOLINES

- This command allows you to extract an isoline curve from a surface, solid, or face of a solid. To issue this command, go to the **Surface** tab, locate the **Curves** panel, and then select the **Extract Isolines** button:

- You will see the following prompts:

```
Select a surface, solid, or face:
Extracting isoline curve in U direction
Select point on surface or [Chain/Direction/Spline points]:
```

- The first line asks you to select a surface, a solid, or a face of a solid. Once this is done, AutoCAD gives you the direction in which the curve will be extracted, U or V (in the above example, the curve is extracted in the U direction). Finally, AutoCAD asks you to specify a point on the surface in order for the extraction process to be finalized.

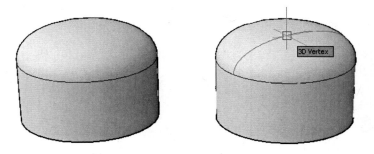

- There are three more options: Chain, which gives a chain of isolines, Direction, which changes the direction of extraction from U to V and vice-versa, and finally, Spline points, which allow you to draw a spline on the curved surface by specifying spline points.
- The objects output from Extract Isolines from Surfaces are splines (even if they look like circles) and from Solids lines, arcs, and circles.

6.12 IMPRINTING SOLIDS

- This command will help you cut an existing face to more faces by using 2D objects as imprints. To issue this command, go to the **Home** tab, locate the **Solid Editing** panel, and then select the **Imprint** button:

- The following prompts will be shown:

```
Select a 3D solid or surface:
Select an object to imprint:
Delete the source object [Yes/No] <N>:
```

- This command will affect both surfaces and solids, which makes it unique. A 2D object should be drawn prior to issuing the command. AutoCAD will ask you to select the desired solid and then select the 2D object. Finally, specify whether to keep the 2D object. The following helps illustrate the concept:

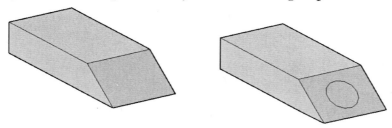

6.13 COLORING SOLID EDGES

- This command allows you to assign colors to selected edges. To issue this command, go to the **Home** tab, locate the **Solid Editing** panel, and then select the **Color Edges** button:

- The following prompts will be shown:

```
Select edges or [Undo/Remove]:
Select edges or [Undo/Remove/ALL]:
```

- Select the desired edges to be colored, and when done press [Enter]. The Select Color dialog box will open. Select the desired color, then click OK.

The colored edges will be shown only in some visual styles, including 2D Wireframe, Hidden, Shaded with Edges, or Wireframe.
- The following helps illustrate the concept:

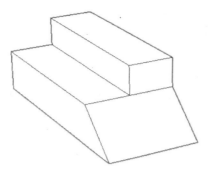

6.14 COPYING SOLID EDGES

- This command allows you to copy the desired edges of a solid to produce a 2D object. This command works like Extract Edges, which we discussed earlier, but this command is different in that it gives you the freedom to select some of the edges instead of all of them. To issue this command, go to the **Home** tab, locate the **Solid Editing** panel, and then select the **Copy Edges** button:

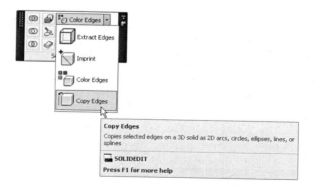

- The following prompts will be shown:

```
Select edges or [Undo/Remove]:
Specify a base point or displacement:
Specify a second point of displacement:
```

- AutoCAD asks you to select the desired edges and then specify a base point and a second point, which is identical to the process of copying objects. When done press [Enter] twice. The following illustrates the concept:

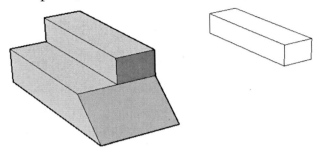

6.15 OFFSETTING SOLID EDGES

- This command allows you to offset a planar solid edge to create a closed polyline or spline using offset distance. To issue this command, go to the **Solid** tab, locate the **Solid Editing**, and then select the **Offset Edge** button:

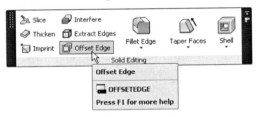

- The following prompts will be shown:

```
Select face:
Specify through point or [Distance/Corner]:
```

- AutoCAD will ask you first to select a face, then to select a through point. This is the default option; specifying this point means you will specify distance and the side of the offset. Another way is to select the Distance option, which will show you the following two prompts:

```
Specify distance <0.0000>:
Specify point on side to offset:
```

- Input the desired distance, and then specify the side of the offset. The last option is Corner, which specifies the type of corners on the offset object when it is created outside the edges of the selected face; there are two choices to select from, either Sharp or Rounded. You can use the output of this command in the Press/Pull command or Extrude command. The following illustrates the concept:

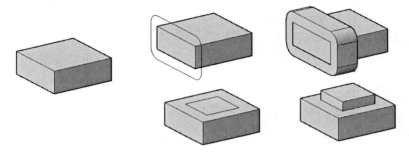

ALL SOLID EDGES FUNCTIONS

Practice 6-5

1. Start AutoCAD 2013.
2. Open **Practice 6-5.dwg**.
3. Using the small rectangle, imprint the big face without keeping the 2D object.
4. Using Extract Faces, extract the newly created face using the polyline as a path. (If you find it hard to select the small face to extrude, use the following trick: select the small face from one of its edges; this way, it will select both faces, then select the Remove option and remove the big face from the selection set.)
5. This is what you should have by now:

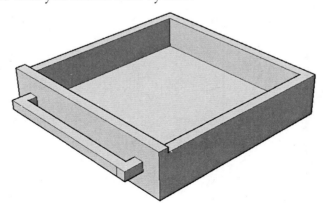

6. Color all the edges red.
7. Change the visual style to Hidden.
8. Using Extracting Edges, extract all the edges of the drawer.
9. Change the visual style to Conceptual.
10. Thaw layer Surface.
11. Using the Extract Isolines command, extract isolines in the U and V directions. Then move the surface away so you can see the isolines alone. What type of objects are the isolines? _____
12. Save and close the file.

6.16 SEPARATING SOLIDS

- Sometimes (because of the Subtract command), you may get two or more non-continuous single solids. This command will rectify this problem by separating the portions so each one of them forms an independent single solid. To issue this command, go to the **Home** tab, locate the **Solid Editing** panel, and then select the **Separate** button:

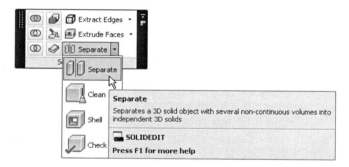

- The following prompts will be shown:

```
Select a 3D solid:
```

- AutoCAD asks you to select the desired solid to separate. When done press [Enter] twice to end the command. The following illustrates the concept:

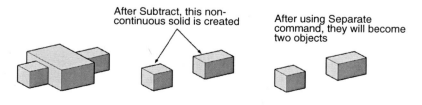

After Subtract, this non-continuous solid is created

After using Separate command, they will become two objects

6.17 SHELLING SOLIDS

- This command will help you create a hollow in a solid. To issue this command, go to the **Home** tab, locate the **Solid Editing** panel, and then select the **Shell** button:

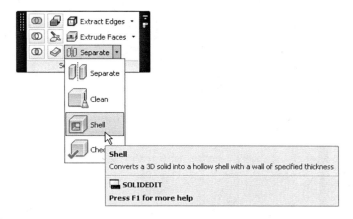

- The following prompts will be shown:

```
Select a 3D solid:
Remove faces or [Undo/Add/ALL]:
Remove faces or [Undo/Add/ALL]:
Enter the shell offset distance:
Solid validation started.
Solid validation completed.
```

- AutoCAD asks you to select the desired solid. This means selecting all its faces, which will create an inside shell, but AutoCAD gives you the ability to remove faces from the shell. AutoCAD will then ask you to input the offset distance. If you input a positive value the outside edge of the solid will stay outside and the offset distance will be to the inside. But if you input a negative value, the outside edge of the solid will be the inside edge and the offset will be to the outside. See the following:

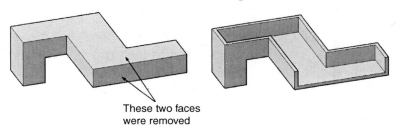

These two faces
were removed

- The other two commands for solids are:
  - Clean, which will help remove any redundant edges/vertices on a solid.
  - Check, which will tell you if the selected object is a solid or any other 3D object.

SEPARATING AND SHELLING A SOLID

Practice 6-6

1. Start AutoCAD 2013.
2. Open **Practice 6-6.dwg**.
3. Using the Shell command, shell the shape to the inside with Offset distance = 0.5.
4. Thaw layer Base.
5. Thaw layer Separate.
6. Using the Subtract command, select the box as the first set, then press [Enter] and select the elliptical shape.
7. Test the resulting shape; is it one object or two objects? _____
8. Using the Separate command, separate the two solids, then test the shapes. What is the result? _____
9. Save and close the file.

6.18 FINDING INTERFERENCE BETWEEN SOLIDS

- This command will help you find (create) the common volume between two sets of solid objects. To issue this command, go to the **Home** tab, locate the **Solid Editing** panel, and then select the **Interfere** button:

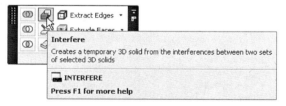

- The following prompts will be shown:

```
Select first set of objects or [Nested selection/Settings]:
Select second set of objects or [Nested selection/checK first
set] <checK>:
```

- This command is done in two steps (with [Enter] between them). You should select the first set of solids, which can be one or more solids, then press [Enter] and then select the second set of solids, which can be one or more solids, and then press [Enter] again. You will see the following dialog box:

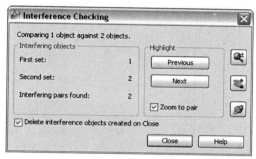

- Under **Interfering objects**, you will find the number of solids in the first and second sets, and finally, the interference pairs found. Use the two buttons labeled **Previous** and **Next** to jump between the solids found as interference solids (if there is a single object, then these two buttons have no effect). If the checkbox Zoom to pair is turned on, AutoCAD will zoom to the interference solids automatically. Next, you can use the three buttons at the right for zooming in and out, panning, and orbiting the interference solid.
- The last step will tell AutoCAD whether or not to keep the interference objects when closing the dialog box.
- While you are selecting your two sets, you will see two options:
  - Nested selection, which allows you to select a nested object within a block or an xref.
  - Settings, which will show the following dialog box:

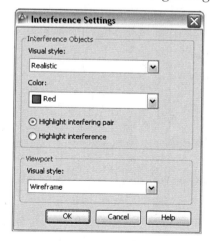

- The above dialog box is easy to understand.
- Note that if you select only the first set without selecting a second set, AutoCAD will find the interference between the solids of the first set. On the other hand, AutoCAD will not try to find solid interference between solids in the same set if there is a single solid in the second set. The following helps clarify the concept:

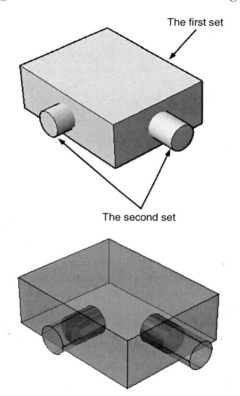

The first set

The second set

FINDING INTERFERENCE BETWEEN SOLIDS

Practice 6-7

1. Start AutoCAD 2013.
2. Open **Practice 6-7.dwg**.
3. Start the Interfere command.
4. Select the box as your first set and the four small cylinders as your second set.
5. Check out the interference volume.

6. Turn off Delete interference objects created on Close, then click Close.
7. Freeze layer Cylinder.
8. Subtract the interference volumes from the box.
9. Thaw layer Cylinder.
10. Using the gizmo move the four cylinders out to the midway.
11. Orbit the shape so you can see that you made a track for the four cylinders.
12. Save and close the file.

NOTES:

CHAPTER REVIEW

1. Which one of the following statements is *not* true?
   a. Offset faces can offset both planar and curved faces.
   b. Extrude faces can offset both planar and curved faces.
   c. Extrude faces can offset only planar faces.
   d. Offset faces can offset only planar faces.
2. In the Shell command, inputting a _____ value means the outside edge of the solid will be the inside edge and the offset will be to the outside.
3. You can offset the faces and edges of a solid.
   a. True
   b. False
4. The difference between the Extract Edges command and the Copying Edges command is:
   a. In the Copying edges command you can't select desired edges.
   b. In the Extract Edges command you can select desired edges.
   c. In the Copying edges command you can select desired edges.
   d. They are the same.
5. In the Interfere command, you have to select two sets, and each set contains a single object.
   a. True
   b. False
6. Use the _____ command when there are two or more non-continuous solid volumes.
7. In the Extrude face command, you can use path with taper angle.
   a. True
   b. False
8. Sometimes AutoCAD will refuse to delete a selected face.
   a. True
   b. False

CHAPTER REVIEW ANSWERS

1. b
3. a
5. b
7. b

Chapter **7** # 3D MODIFYING COMMANDS

In This Chapter

◇ 3D Move, 3D Rotate, and 3D Scale
◇ 3D Array and 3D Align
◇ Filleting and chamfering solids

7.1 3D MOVE, 3D ROTATE, AND 3D SCALE AND SUBOBJECTS AND GIZMOS

- In Chapters 2 and 3 of this book we discussed selecting subobjects (vertices, edges, and faces) for both solids and meshes. Then we discussed gizmos, which allow you to perform three commands: Move, Rotate, and Scale in the 3D space. In the older versions of AutoCAD we used other commands, such as 3D Move, 3D Rotate, and 3D Scale, but subobjects and gizmos are identical to these three commands. In this section we will go over the three commands, just for the sake of knowing them, but we highly recommend using subobjects and gizmos instead.

7.1.1 3D Move Command

- To issue this command, go to the **Home** tab, locate the **Modify** panel, and then select the **3D Move** button:

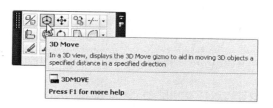

- The following prompts will be shown:

```
Select objects:
Specify base point or [Displacement] <Displacement>:
Specify second point or <use first point as displacement>:
```

- Select the desired objects or subobjects and perform moving in 3D.

7.1.2 3D Rotate Command

- To issue this command, go to the **Home** tab, locate the **Modify** panel, and then select the **3D Rotate** button:

- The following prompts will be shown:

```
Select objects:
Specify base point:
Pick a rotation axis:
Specify angle start point or type an angle:
```

- Select the desired objects or subobjects and perform rotating in 3D.

7.1.3 3D Scale Command

- To issue this command, go to the **Home** tab, locate the **Modify** panel, and then select the **3D Scale** button:

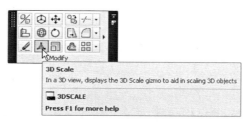

- The following prompts will be shown:

```
Select objects:
Specify base point:
```

```
Pick a scale axis or plane:
Specify scale factor or [Copy/Reference] <1.0000>:
```

- Select the desired objects or subobjects and perform scaling in 3D.

7.2 MIRRORING IN 3D

- This command allows you to create a mirror image in the 3D space using a mirroring plane. This is different than the normal Mirror command, which works only in the current XY plane. To issue this command, go to the **Home** tab, locate the **Modify** panel, and select the **3D Mirror** button:

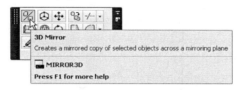

- The following prompts will be shown:

```
Select objects:
Select objects:
Specify first point of mirror plane (3 points) or
[Object/Last/Zaxis/View/XY/YZ/ZX/3points] <3points>:
```

- AutoCAD offers multiple methods to create a mirroring plane, and some of them are identical to the UCS options. They are discussed as follows.

7.2.1 XY, YZ, and ZX Options

- We put these three options together because they are similar. These options will help you create a mirroring plane parallel to the original three planes. You will see the following prompts:

```
[Object/Last/Zaxis/View/XY/YZ/ZX/3points] <3points>:
Specify point on ZX plane <0,0,0>:
Delete source objects? [Yes/No] <N>:
```

- In this example we used the ZX option. AutoCAD will ask you to specify a point on the selected plane; specify your point.

7.2.2 3 Points Option

- To create any plane, you should specify 3 points; this option is identical to the one in the UCS command. This option is also considered the most generic option among all other options, since it gives you the freedom to create any mirroring plane. The following prompts will be shown:

```
[Object/Last/Zaxis/View/XY/YZ/ZX/3points] <3points>:
Specify first point on mirror plane:
Specify second point on mirror plane:
Specify third point on mirror plane:
Delete source objects? [Yes/No] <N>:
```

7.2.3 Last Option

- This option will help you retrieve the last used mirroring plane used.

7.2.4 Object, Zaxis, View Options

- All of these options were discussed with the UCS command.
- For all the above options, the last prompt will be whether to delete the source object.

MIRRORING IN 3D

Practice 7-1

1. Start AutoCAD 2013.
2. Open **Practice 7-1.dwg**.
3. Using 3D Mirror command, mirror the shape to complete it.
4. Union the two shapes.
5. You should get the following shape:

6. Save and close the file.

7.3 ARRAYING IN 3D

- The same command you use for 2D arraying will be used for 3D arraying after changing some values to give it the 3D depth. There are three different types of arrays in AutoCAD:
  - Rectangular Array
  - Path Array
  - Polar Array
- These three commands are associative and allow you to edit the values of the array even after the command is finished. To issue these commands, go to the **Home** tab, locate the **Modify** panel, and select one of the three commands:

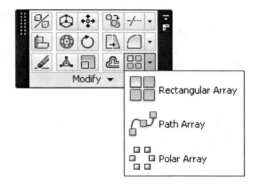

- After you finish creating any normal 2D associative array (Rectangular, Path, and Polar), click it, and a new context tab named **Array** will appear:

- In this context tab there are two places to make your rectangular array act as a 3D array. The first place is the **Levels** panel, which allows you to create the third dimension of the array. Input the following:
  - Level Count
  - Level Spacing
  - Total Level Distance

- The second place is the **Rows** panel. After extending the panel, you will see the following:

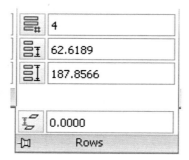

- This piece of information is the **Incremental Elevation,** which allows you to give the second row an elevation similar to the first row, then the third row will double the elevation, and so on. The following shows the difference between levels and incremental elevation (the example here is a rectangular array):

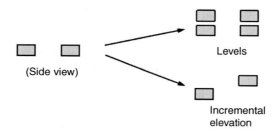

(Side view)

Levels

Incremental
elevation

ARRAYING IN 3D

Practice 7-2

1. Start AutoCAD 2013.
2. Open **Practice 7-2.dwg**.
3. Using Rectangular Array, array the chair at the left (near the UCS icon) using the following information: No. of rows = 3, No. of columns = 10, Row Spacing = 27.5, and Column Spacing = 20, Incremental Elevation = 6.
4. Freeze layer Curved.
5. Using Path Array, array the chair at the beginning of the curved stage using the following information: the path curve is the small arc, No. of items = 16,

Divide evenly along path, No. of rows = 3, Row Spacing = 27.5, Incremental Elevation = 6.

6. Thaw layer Curved.
7. If you have time, do the rest of the chairs.
8. The output should look like the following:

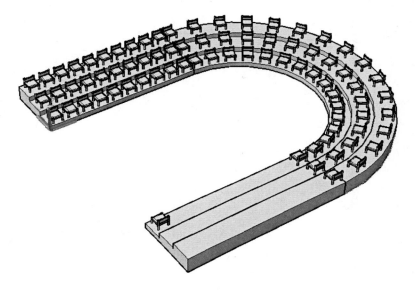

9. Save and close the file.

7.4 ALIGNING OBJECTS IN 3D

■ This command allows you to align 3D objects. Aligning 3D objects involves moving, rotating, and scaling in some cases. You will use one, two, or three source points to align them to one, two, or three destination points. To issue this command, go to the **Home** tab, locate the **Modify** panel, and select the **3D Align** button:

- The following prompts will be shown:

```
Select objects:
Specify source plane and orientation ...
Specify base point or [Copy]:
Specify second point or [Continue] <C>:
Specify third point or [Continue] <C>:
Specify destination plane and orientation ...
Specify first destination point:
Specify second destination point or [eXit] <X>:
Specify third destination point or [eXit] <X>:
```

- The first step is to select the objects to be aligned, then specify one, two, or three points of the source plane. After the first point you can use the **Continue** option, which allows you to jump to the destination plane. After the first point you can also specify whether to use a **copy** of the original object to keep the original object intact. The last step is to specify one, two, or three points as destination points; after the first point you can use the **exit** option to end the command.

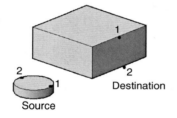 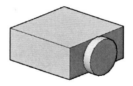

7.5 FILLETING SOLID EDGES

- This command allows you to fillet solid edges. Although you can still use the 2D Fillet command, we highly recommend using this command instead. To issue this command, go to the **Solid** tab, locate the **Solid Editing** panel, and then select the **Fillet Edge** button:

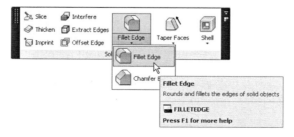

- The following prompts will be shown:

```
Radius = 1.0000
Select an edge or [Chain/Loop/Radius]:
Select an edge or [Chain/Loop/Radius]:
2 edge(s) selected for fillet.
```

- The first line reports the current value of Radius. AutoCAD then asks you to select the desired edges; if you don't change the radius value, it will use the default value. Once you are done selecting edges, press [Enter] and AutoCAD will report the number of edges selected. The following message will be shown:

```
Press Enter to accept the fillet or [Radius]:
```

- Either press [Enter] to accept the fillet produced or edit the Radius value. You can do this by either moving the arrow representing the radius value, or by selecting the **Radius** option and typing in the new value:

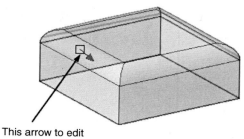

This arrow to edit
Radius value

7.5.1 Chain Option

- Using this option you can select more than one edge if the edges are tangent to one another. When you select this option, you will see the following prompt:

```
Select an edge chain or [Edge/Radius]:
```

- The condition here is tangency; see the following example, which helps to illustrate the **Chain** option:

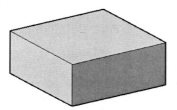

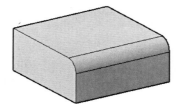

Edges are not tangent to each   Using Chain option, one edge was filleted

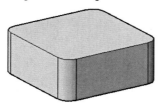

Edges are tangent to each other   Using Chain option, all edge was filleted

7.5.2 Loop Option

- This option allows you to select a loop of edges regardless of if they are tangent to each other or not. There are two possible loops for any given edge; AutoCAD will randomly select one of these two loops, asking you either to accept it or to select the next loop. The following shows the first step after selecting an edge:

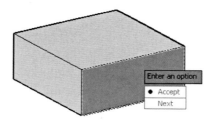

- After selecting the Next option, you will see that all edges are filleted in one step:

7.6 CHAMFERING SOLID EDGES

- This command allows you to chamfer selected solid edges using distances. To issue this command, go to the **Solid** tab, locate the **Solid Editing** panel, and then select the **Chamfer Edge** button:

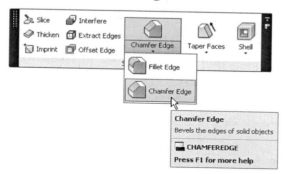

- The following prompts will be shown:

```
Distance1 = 1.0000, Distance2 = 1.0000
Select an edge or [Loop/Distance]:
Select an edge belongs to the same face or [Loop/Distance]:
Select an edge belongs to the same face or [Loop/Distance]:
```

- The first message reports the current values of the two distances. AutoCAD then asks you to select the desired edges; if you don't change the distance values, AutoCAD will use the default value. Once you are done selecting edges, press [Enter] and the following message will be shown:

```
Press Enter to accept the chamfer or [Distance]:
```

- Either press [Enter] to accept the chamfer produced or edit the **Distance** values. You can do this by either moving the two arrows representing the chamfer distance values or by selecting the **Distance** option and typing in two new values. The following helps illustrate the concept:

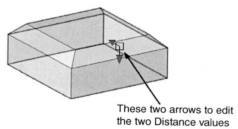

These two arrows to edit
the two Distance values

- The Loop option is identical to the one in the Fillet Edge command, and we will not discuss it again.

ALIGNING, FILLETING, AND CHAMFERING

Practice 7-3

1. Start AutoCAD 2013.
2. Open **Practice 7-3.dwg**.
3. Using Align 3D, align the cylinder to the center of the box as follows:

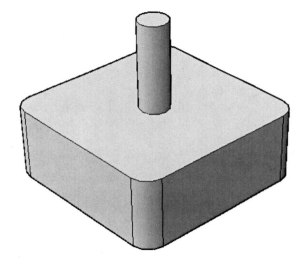

4. Look at the model using Top view and Parallel.
5. Move the cylinder 16 units to the right, then 16 units up.
6. Copy the cylinder to the four edges of the box using 32 units in all directions.
7. Using the gizmo and the Move command, move the four cylinders down by 25 units.
8. Subtract the four cylinders from the big shape.
9. Using the Fillet Edge command and the Chain option, fillet the upper and lower edges using radius = 2.5 units.
10. Using the Chamfer Edge command and distance 1 = distance 2 = 2 units, chamfer the top edges of the four cylinders.

11. The final shape should look like the following:

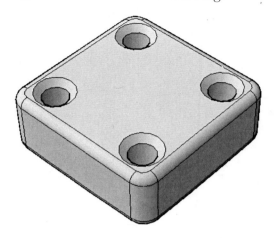

12. Save and close the file.

PROJECT USING METRIC UNITS

Project 7-1-M

1. Start AutoCAD 2013.
2. Open **Project 7-1-M.dwg**.
3. Using Taper Faces, taper the two faces by angle = 15° to look like the following:

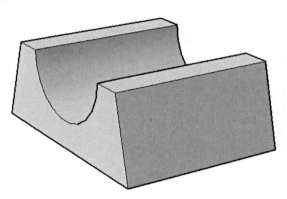

4. Thaw layer Path.

5. Extrude the front face using the path, and you should get the following shape:

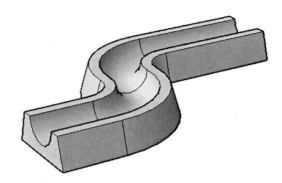

6. Using Mirror 3D, mirror the shape using ZX plane as the mirroring plane.
7. Union the two shapes.
8. Thaw layers Circle and Path.
9. Using the Extrude command, extrude the circle using the path.
10. Move the pipes to the center of the base.
11. Shell the pipes to the inside by 6 units.
12. You should get the following shape:

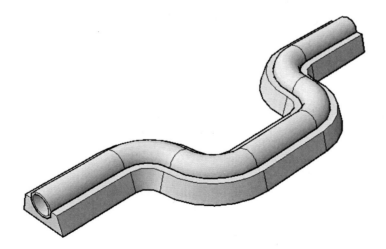

13. Using the Quadrant OSNAP, draw a line as shown:

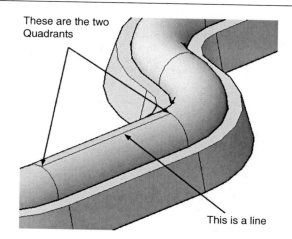

These are the two Quadrants

This is a line

14. Using the two ends and the midpoint of the line, draw three circles with R = 32 units.
15. Erase the line.
16. Using the Press/Pull command, press the three circles to the inside by 32 units, then subtract the new shapes from the pipe.
17. Using the Extrude command, extrude the two circles at the two ends by 260 units up.
18. As for the hole at the middle, it appears that we don't need it so delete the face, along with the circle.
19. Using the gizmo move the two cylinders downward by 20 units.
20. Using Offset Edge, offset the two top faces of the two newly created cylinders to the inside using distance = 6 units.
21. Using Press/Pull, press the two circles by 260 units to shell the two cylinders.
22. The shape should look like the following:

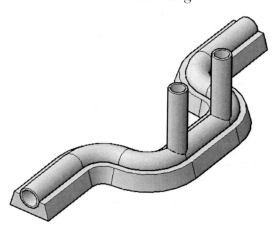

23. Draw a line from the edge as shown below in the positive X direction (WCS) of distance = 400 units:

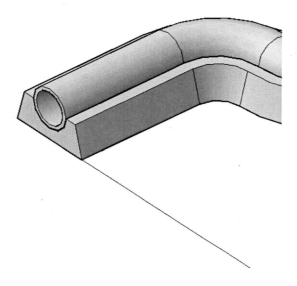

24. Using the Polyline command, draw the following shape at the edge of the line (all the lines are 8 units, except the long ones, which are 24 units):

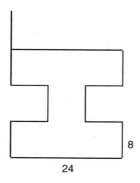

8

24

25. Using the Extrude command, extrude the polyline up by 480 units.
26. Using 3D Array (Rectangular), create an array with the following information:
    a. Number of rows = 4
    b. Number of columns = 2
    c. Number of levels = 2
    d. Row spacing = 400
    e. Column spacing = –800
    f. Level spacing = 510

27. Draw a box 824x1224x30 and copy it to cover the first set of columns, and copy it again to cover the second set of columns.
28. Fillet the edge as shown below using R = 12 units.

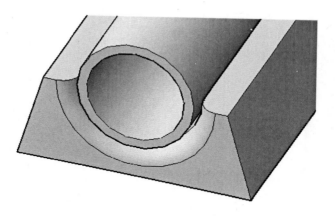

29. Do the same thing for the other end.
30. Using the Fillet Edge command and R = 6 and the Chain option, fillet the edge as shown:

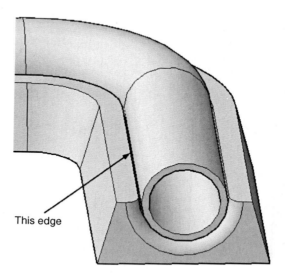

This edge

31. Do the same thing for the other edges.
32. Save and close the file.

PROJECT USING IMPERIAL UNITS

Project 7-1-I

1. Start AutoCAD 2013.
2. Open **Project 7-1-I.dwg**.
3. Using Taper Faces, taper the two faces by angle = 15° to look like the following:

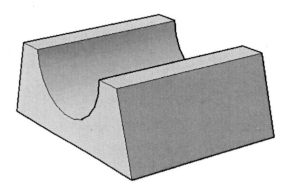

4. Thaw layer Path.
5. Extrude the front face using the path, and you should get the following shape:

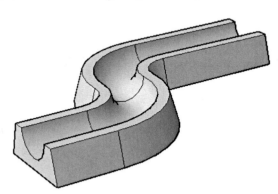

6. Using Mirror 3D, mirror the shape using the ZX plane as the mirroring plane.
7. Union the two shapes.
8. Thaw layers Circle and Path.
9. Using the Extrude command, extrude the circle using the path.

10. Move the pipes to the center of the base.
11. Shell the pipes to the inside by 3".
12. You should get the following shape:

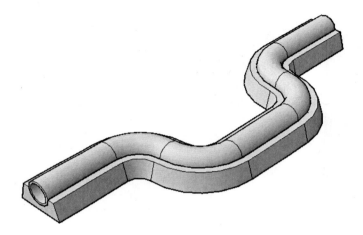

13. Using the Quadrant OSNAP, draw a line as shown:

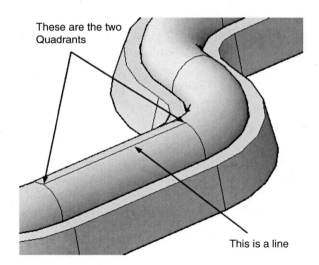

These are the two Quadrants

This is a line

14. Using the two ends and the midpoint of the line, draw three circles with R=13".
15. Erase the line.
16. Using the Press/Pull command, press the three circles to the inside by 13", then subtract the new shapes from the pipe.

17. Using the Extrude command, extrude the two circles at the two ends by 8'-8" up.
18. As for the hole at the middle, it appears that we don't need it so delete the face, along with the circle.
19. Using the gizmo move the two cylinders downward by 8".
20. Using the Offset Edge, offset the two top faces of the two newly created cylinders to the inside using distance = 3".
21. Using Press/Pull, press the two circles by 8'-8" to shell the two cylinders.
22. The shape should look like the following:

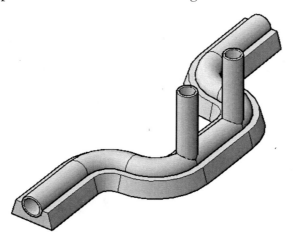

23. Draw a line from the edge as shown below in the positive X direction (WCS) of distance = 14':

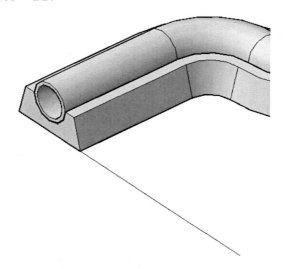

24. Using the Polyline command, draw the following shape at the edge of the line (all the lines are 3", except the long ones, which are 10"):

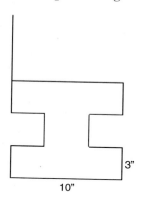

25. Using the Extrude command, extrude the polyline up by 16'.
26. Using 3D Array (rectangular), create an array with the following information:
    a. Number of rows = 4
    b. Number of columns = 2
    c. Number of levels = 2
    d. Row spacing = 14'
    e. Column spacing = −28'
    f. Level spacing = 17'
27. Draw a box 28'-9.0"x42'-10.0"x1'-0" and copy it to cover the first set of columns, and copy it again to cover the second set of columns.
28. Fillet the edge as shown below using R = 5":

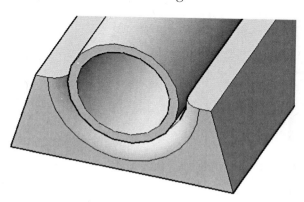

29. Do the same thing for the other end.

30. Using the Fillet Edge command and R = 3" and the Chain option, fillet the edge as shown:

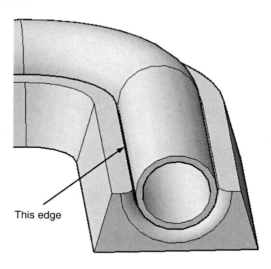

This edge

31. Do the same thing for the other edges.
32. Save and close the file.

NOTES:

CHAPTER REVIEW

1. You can delete a fillet of a solid.
   a. True
   b. False
2. The difference between Extrude Face and Offset is:
   a. Using Extrude you can extrude a curved face.
   b. Using Offset you can't offset a curved face.
   c. Offset can offset both planar and curved faces.
   d. Extrude face can extrude both planar and curved faces.
3. Using the Array command in the 3D environment you can set _____ and _____ to utilize the 3D feature of this command.
4. In the Shell command, inputting a negative value means:
   a. The outer face will be the inner face.
   b. The inner face will be the outer face.
   c. Positive and negative values are the same.
   d. The Shell command will not accept negative values.
5. For the Chain option in the Fillet Edge command to be successful the edges should be tangent to each other.
   a. True
   b. False
6. For the Loop option in the Chamfer Edge command to be successful the edges should be tangent to each other.
   a. True
   b. False

CHAPTER REVIEW ANSWERS

1. a
3. Levels, Incremental Elevation
5. a

8

CONVERTING
AND SECTIONING

In This Chapter
◇ Converting objects
◇ Sectioning using the Slice, Section Plane, and Flatshot commands

8.1 INTRODUCTION TO CONVERTING 3D OBJECTS

- In the previous chapters we outlined the different types of 3D objects in AutoCAD: solids, meshes, and surfaces. While we were discussing these types we touched on some techniques to convert some objects (whether 2D or 3D) to solids, meshes, or surfaces. The techniques we discussed were:
  - Converting 2D objects to solids or surfaces using the Extrude, Loft, Revolve, Sweep commands.
  - Converting 2D objects to solids using the Press/Pull command.
  - Converting solids to meshes using the Smooth command.
  - Converting solids, meshes, and surfaces to the NURBS surface using the Convert to NURBS command.
  - Converting surfaces to solids using the Sculpt command.
- In this chapter, we will cover more techniques for converting objects when producing a 3D drawing in AutoCAD.

8.2 CONVERTING 2D AND 3D OBJECTS TO SOLIDS

- In this section we will discuss the Thicken command, which will convert any surface to a solid, as well as the Convert to Solid command, a generic command that can convert meshes and some 2D objects to solids.

8.2.1 Thicken Command

- This command allows you to convert any surface to solids. To issue this command, go to the **Home** tab, locate the **Solid Editing** panel, and then select the **Thicken** button:

- The following prompts will be shown:

```
Select surfaces to thicken:
Select surfaces to thicken:
Specify thickness:
```

- AutoCAD will ask you to select the desired surface; you can select any number of surfaces. When you are done press [Enter] and input the value of the thickness. The following illustrates the concept:

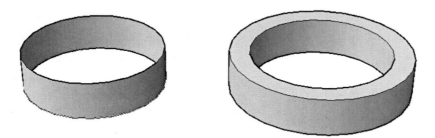

8.2.3 Convert to Solid Command

- This command allows you to convert closed and non-intersecting meshes or closed 2D objects with thickness value. To issue this command, go to the **Home** tab, locate the **Solid Editing** panel, and then select the **Convert to Solid** button:

- The following prompts will be shown:

```
Mesh conversion set to: Smooth and optimized.
Select objects:
Select objects:
```

- The first line gives you the current settings of smoothness. Then AutoCAD will ask you to select a mesh or closed 2D object (circle, ellipse, or closed polyline) with thickness. When done press [Enter]. See the following illustration:

- In order to control the level of smoothness of the conversion process, you should go to the **Mesh Modeling** tab, locate the **Convert Mesh** panel, and using the button at the right as illustrated below, choose from four values:
    - Smooth, optimized
    - Smooth, not optimized
    - Faceted, optimized
    - Faceted, not optimized

- They are self-explanatory:

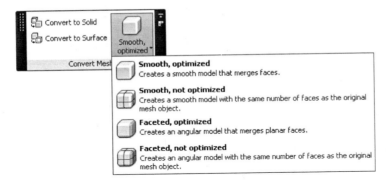

- See the following example:

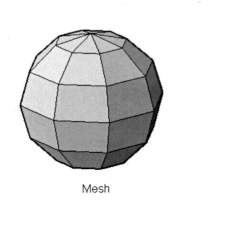

Mesh

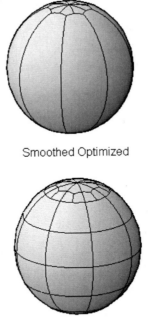

Smoothed Optimized

Smoothed Not Optimized

- Notice how the faces were merged in "Smoothed Optimized." Compare them with the result of "Smooth Not Optimized."

The meaning of a 2D object with thickness is illustrated below:

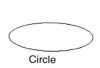

Circle   Circle with Thickness

- If you draw a closed polyline, circle, or ellipse, then using the **Properties** palette assign a value for *Thickness*; then you can convert it to a solid. The following illustrates the concept:

8.3 CONVERTING 2D AND 3D OBJECTS TO SURFACES

- The following two commands allow you to convert solids and meshes to surfaces. These commands are:
  - Convert to Surface
  - Explode
- We will discuss these two commands in the following sections.

8.3.1 Convert to Surface Command

- This command will convert 3D objects such as solids and meshes to surfaces. It can also convert lines, polylines, circles, arcs, with thickness, or regions to surfaces. To issue this command, go to the **Home** tab, locate the **Solid Editing** panel, and then select the **Convert to Surface** button:

- The following prompts will be shown:

```
Mesh conversion set to: Smooth and optimized.
Select objects:
Select objects:
```

- The first line gives you the current settings of smoothness. Then AutoCAD will ask you to select objects. When done press [Enter]. See the following illustration:

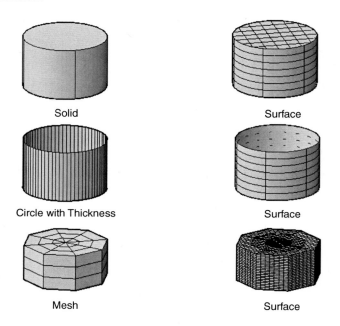

Solid	Surface
Circle with Thickness	Surface
Mesh	Surface

8.3.2 Explode Command

- This command can be used with some solid objects. The following are the solid objects that can converted to surfaces using the Explode command:
  - Cylinder
  - Cone
  - Sphere
  - Torus
- The top and bottom of the cylinder will be converted to a region and the body will be converted to a surface. As for the cone, the bottom will be converted to a region and the body will be converted to a surface. The following illustrates the concept:

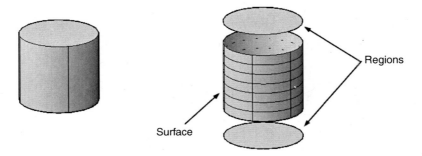

Surface

Regions

8.4 CONVERSION BETWEEN OBJECTS – CONCLUSION

- In conclusion, see the following table, which outlines all the commands that can be used to convert an object to another:

	From 2D Objects	*From Mesh*	*From Solid*	*From Surface*
To Solid	Extrude, Loft, Revolve, Sweep, Press/Pull	Convert to Solid	--	Sculpt, Thicken, Convert to Solid
To Surface	Extrude, Loft, Revolve, Sweep	Convert to Surface, Convert to NURBS	Convert to Surface, Convert to NURBS, Explode (some objects only)	--
To Mesh	--	--	Smooth	Smooth

CONVERTING OBJECTS

Practice 8-1

1. Start AutoCAD 2013.
2. Open **Practice 8-1.dwg**.
3. You will see a mesh sphere. Go to the Mesh tab, locate the Convert Mesh panel, select Smoothed Not Optimized, and then select the Convert to Solid command.
4. Undo what you did.
5. Select Smoothed Optimized, then select the Convert to Solid command.

6. Compare the number of faces in both cases. Which has more faces? _____ Why? _____

7. Freeze layer Mesh and Thaw layer Surface.

8. Set VSEDGES to 1.

9. Use the Thicken command with thickness = 5.

10. Change the visual style to X-Ray. Do you see a hollow sphere with thickness?

11. Change the visual style to Conceptual.

12. Freeze layer Surface and Thaw layer Solid.

13. Explode the solid.

14. What is the result? _____

15. Save and close the file.

8.5 INTRODUCTION TO SECTIONING IN 3D

- In this section we will discuss three commands, and all of them allow you to produce a section of the 3D model:
  - The Slice command, which will work on solids and surfaces, but not on meshes.
  - The Section Plane command, which will work on all types of 3D objects.
  - The Flatshot command, which will create a shot parallel to the current view of any 3D object in the drawing.
- Each of these commands is unique and will give you the ability to create a 2D section of the 3D shape using different methods.

8.6 SECTIONING USING THE SLICE COMMAND

- This command allows you to create a 3D slice of the 3D object using the slicing plane. To issue this command, go to the **Home** tab, locate the **Solid Editing** panel, and then select the **Slice** button:

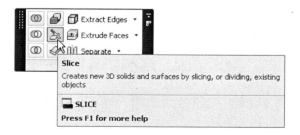

- The following prompts will be shown:

```
Select objects to slice:
Specify start point of slicing plane or [planar
Object/Surface/Zaxis/View/XY/YZ/ZX/3points] <3points>:
Specify a point on the ZX-plane <0,0,0>:
Specify a point on desired side or [keep Both sides] <Both>:
```

- AutoCAD wants you to select the desired solid or a surface, then specify a slicing plane. These methods were all discussed in Chapter 7 with the Mirror 3D command, except the Surface option, which will align the slicing plane to a surface (Revsurf, Tabsurf, Rulesurf, and Edgesurf). Finally, AutoCAD wants you to specify a point on the desired side or you can choose to keep both sides.

SLICING OBJECTS

Practice 8-2

1. Start AutoCAD 2013.
2. Open **Practice 8-2.dwg**.
3. Use the Slice command to get the following result:

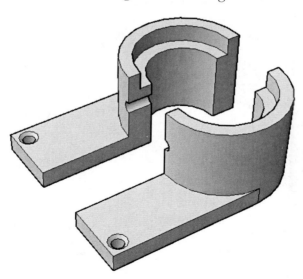

4. Save and close the file.

8.7 SECTIONING USING SECTION PLANE COMMAND

- This command allows you to create 2D/3D section from any 3D object in the current drawing. It also has the ability to insert it as a block or export it to another file. To issue this command, go to the **Home** tab, locate the **Section** panel, and then select the **Section Plane** button:

- The following prompts will be shown:

```
Select face or any point to locate section line or
[Draw section/Orthographic]:
```

- There are three methods to create a section plane; they are discussed in the following sections.

8.7.1 Select a Face Method

- This is the default option. This option will ask you to select one of the faces to use as a section plane. The following illustrates the concept:

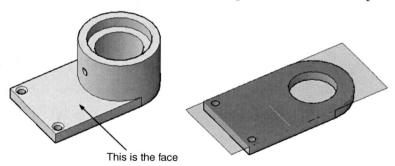

This is the face

8.7.2 Draw Section Option

- Using this option you will be able to draw a section plane using OSNAPs. The following prompts will be shown:

```
Specify start point:
Specify next point:
Specify next point or ENTER to complete:
Specify next point or ENTER to complete:
Specify point in direction of section view:
```

- AutoCAD asks you to specify points in the same manner as drafting a line or polyline, and you will press [Enter] when you are done. AutoCAD needs a minimum of two points to draw a section plane. Finally, click the side you want the sectioning to work at. The following illustrates the concept:

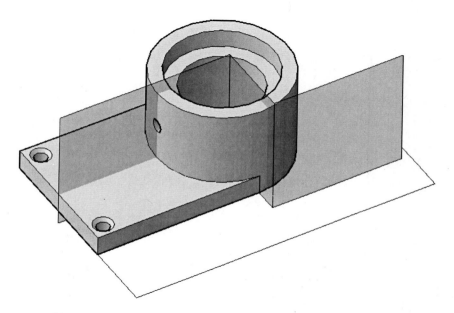

8.7.3 Orthographic Option

- Using this option you will be able to create a section plane parallel to one of the six orthographic planes relative to the current UCS. The following prompts will be shown:

```
Align section to: [Front/bAck/Top/Bottom/Left/Right]:
```

- Select the desired plane, and the section plane will be implemented on the 3D object.

- The following illustrates the concept:

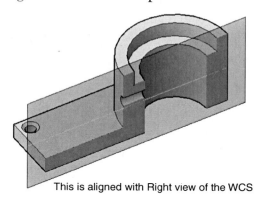

This is aligned with Right view of the WCS

8.7.4 Section Options

- After creating a section plane, you should click the plane to check the three options available. You will see a triangle as shown below, and when clicked you will see the following:

- You will see three section options:
  - Section Plane (default option), which will show you only the section plane. See the following:

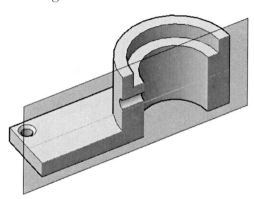

- Section Boundary, which allows you to set the 2D range that controls what will be affected by the section plane. See the following:

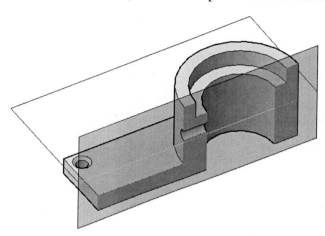

- Section Volume, which allows you to set the 3D range to control the objects in 3D that are affected by the section plane. See the following:

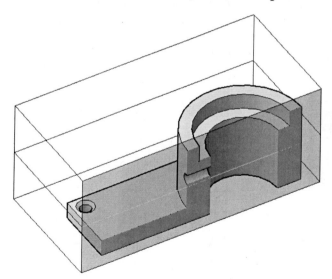

8.7.5 Section Plane Grips

- After creating and controlling the section plane you can further manipulate the section plane using grips. If you click the section plane you will see something like the following:

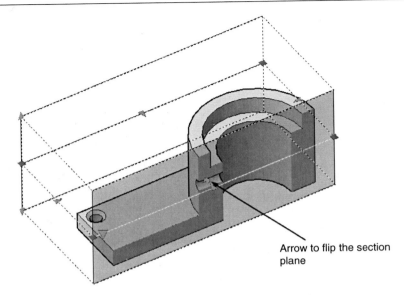

Arrow to flip the section
plane

- You can use the grips, the rectangles, and the triangle grips to control the size
  of the volume of the section plane. You can use the arrow to flip the side of
  the section.

8.7.6 Adding a Jog to an Existing Section Plane

- Using this command you can add a jog to an existing section plane (you can
  still draw a jog using the Draw Section option discussed above). To issue this
  command, go to the **Home** tab, locate the **Section** panel, and then select
  the **Add Jog** button:

- The following prompts will be shown:

```
Select section object:
Specify a point on the section line to add jog:
```

- AutoCAD will ask you to specify a point on the section line and will add a jog instantly. In each command, you can add a single jog, or as many jogs as you want. See the following:

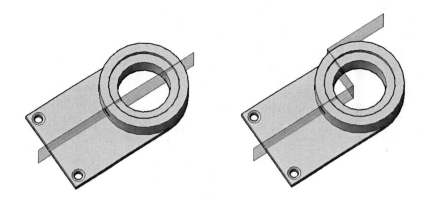

- After adding the jog, you can modify it using the grips. See the following:

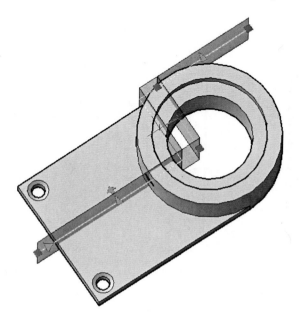

8.7.7 Live Section Command

- This command allows you to activate the section plane. To issue this command, go to the **Home** tab, locate the **Section** panel, and then select the **Live Section** button:

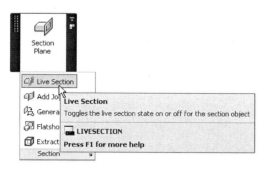

- Another method to invoke this command is to select the section plane and right-click, and then select the **Activate live sectioning** option:

- This is what happens after activating a section plane:

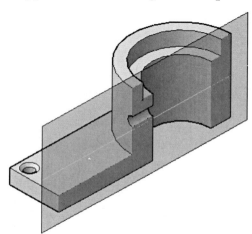

- Using right-click, you can select to **Show cut-away geometry**, and the following will be shown:

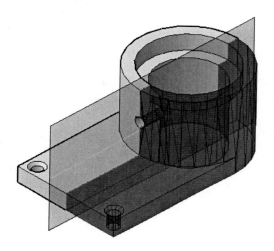

8.7.8 Generate Section Command

- This command allows you to create a 2D/3D section and insert it as a block in your drawing or export it to an external file. To issue this command, go to **Home** tab, locate the **Section** panel, and then select the **Generate Section** button:

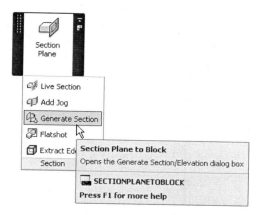

- Another way to reach this command is by selecting the section plane and right-clicking, and then selecting the **Generate 2D/3D section** option:

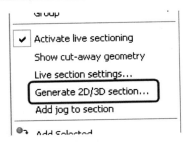

- The following dialog box will be shown:

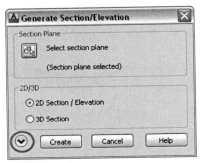

- If you click the arrow pointing down, you will see the following:

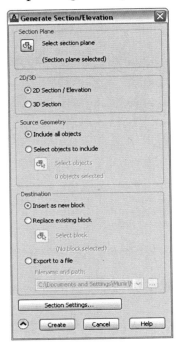

- The first button at the top allows you to select the desired section plane, then select whether you want to create a **2D Section/Elevation** or a **3D Section**. After selecting, click the **Create** button, and the following prompts will be shown:

```
Units: Inches Conversion: 1.0000
Specify insertion point or [Basepoint/Scale/X/Y/Z/Rotate]:
Enter X scale factor, specify opposite corner, or
[Corner/XYZ] <1>:
Enter Y scale factor <use X scale factor>:
Specify rotation angle <0>:
```

- These prompts are the same as the Insert Block command prompts, so we will not discuss them.
- The following is an example of a 2D section (notice the dashed line to indicate a hidden line):

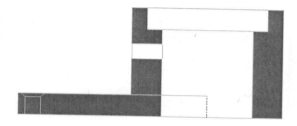

- The following is an example of a 3D section:

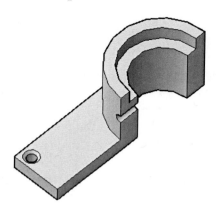

- You can change all or any of the following:
  - Whether to include all objects, or select part of the drawing to be included in the sectioning process

- Whether to insert the section as a new block, to update an existing block (sectioning process took place in this drawing before), or to export the section to DWG file
- The Section Settings; if you click this button you will see the following dialog box:

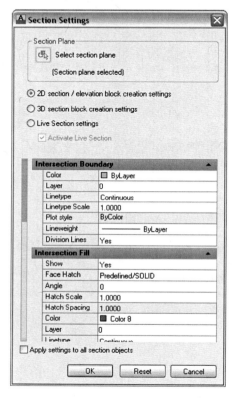

- This dialog box will control the settings for 2D sections, 3D sections, and Live sections. You can control the Color, Linetype, Linetype Scale, and Lineweight of things such as Intersection Boundary, Intersection Fill, Boundary Lines, and Cut-away Geometry.

USING THE SECTION PLANE COMMAND

 Practice 8-3

1. Start AutoCAD 2013.
2. Open **Practice 8-3.dwg**.

3. Using the Section Plane command, select the following face:

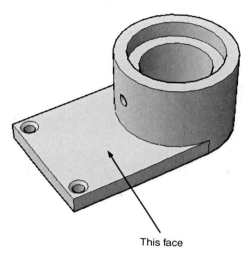

This face

4. Using the Move Gizmo, move the section plane in the Z direction up 13.5 units.
5. Generate a 2D section and insert it as a block in the drawing, without changing the settings.
6. Erase the section plane.
7. Create another orthographic section plane aligned with Right.
8. Generate a 2D section, but this time change the Section Settings for Intersection Fill for the Face Hatch to ANSI31, with Scale = 5, and export it to a file with the name *Section Right.dwg*.
9. Erase the section plane.
10. Create a section plane using the Draw Section option like the following:

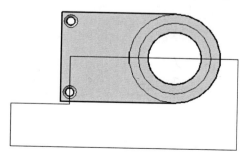

11. Generate a 3D section and export to a file with the name *Section 3D.dwg*.
12. Save and close the file.

8.8 SECTIONING USING THE FLATSHOT COMMAND

- This command allows you to create a 2D block representing one of the views of the 3D model using the ViewCube (or any other viewing command).
- To issue this command, go to the **Home** tab, locate the **Section** panel, and then select the **Flatshot** button:

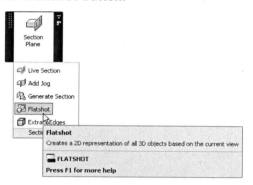

- Before issuing the command note the following two points:
  - Choose the desired view (top, bottom, right, left, etc.).
  - To get the right section make sure you are in *Parallel* mode and not in *Perspective* mode. You can do that using the ViewCube.
- The following dialog box will be displayed:

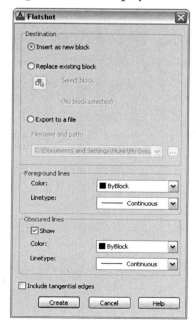

- The upper part of the dialog box is similar to the one discussed in the Section Plane command section. In the rest of the dialog box:
  - Specify the Color and Linetype of the foreground lines.
  - Select whether to Show/Hide the Obscured lines, then specify the Color and Linetype for them (make sure you change them to Dashed to simulate the right drafting standards).
- When done click the **Create** button, and the following prompts will be shown:

```
Units: Inches Conversion: 1.0000
Specify insertion point or [Basepoint/Scale/X/Y/Z/Rotate]:
Enter X scale factor, specify opposite corner, or
[Corner/XYZ] <1>:
Enter Y scale factor <use X scale factor>:
Specify rotation angle <0>:
```

- These prompts are similar to the ones for the Insert Block command.
- The following is an example of the Flatshot command. Here is the top of a 3D shape produced by the Flatshot command:

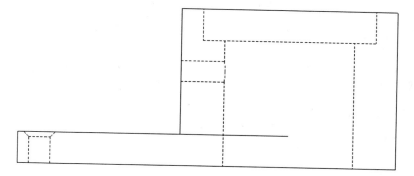

USING THE FLATSHOT COMMAND

Practice 8-4

1. Start AutoCAD 2013.
2. Open **Practice 8-4.dwg**.
3. Look at the top and right views and generate two flatshots, and don't forget to change the linetype of the obscured lines to Dashed.
4. Save and close the file.

NOTES:

CHAPTER REVIEW

1. Thicken will convert a mesh to a solid.
   a. True
   b. False
2. In order to produce the right results using the Flatshot command, you should turn the view to _____ and not _____ .
3. Which of the following is *not* a method for creating a section plane?
   a. Selecting a face
   b. Draw Section
   c. Parallel to one of the XY, YZ, and ZX planes
   d. Orthographic
4. Convert to Solid will convert both meshes and surfaces to solids.
   a. True
   b. False
5. Which one of the following is *not* a method for creating a slice plane?
   a. Planar object
   b. 3 points
   c. Orthographic
   d. Surface
6. The _____ command will convert both solids and surfaces to meshes.

CHAPTER REVIEW ANSWERS

1. b
3. c
5. c

9

PRINTING IN 3D AND CREATING 3D DWF FILES

In This Chapter

◇ Named views and viewports in 3D
◇ Model Documentation
◇ Generating and viewing 3D DWF files

9.1 NAMED VIEWS AND 3D

- You will use the 3D viewing commands to look at your model from different angles. While using these commands you can save time and name different views for use later or for use in layouts for printing purposes. To issue this command, go to the **View** tab, locate the **Views** panel, and then select the **View Manager** button:

- The following dialog box will be shown:

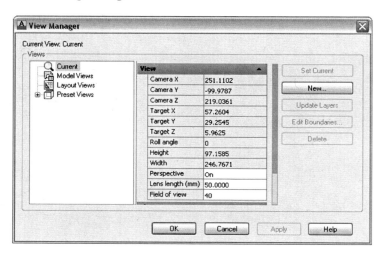

- To create a new view click the **New** button, and the following dialog box will appear:

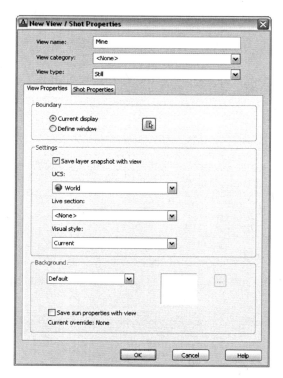

- First type in the View's name. You can save time by setting the following:
  - The current layer status.
  - The current UCS.
  - The current Live section.
  - The current Visual style.
- You can also add a background to the saved view; choose one of five choices:

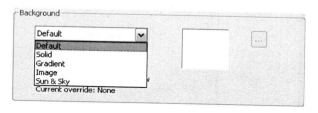

- A dialog box will pop up where you can manipulate the relative parameters of each one of the five selections. Once you are done, click OK to end the command.
- You will see the following picture. In this dialog box you will see a list of the saved views

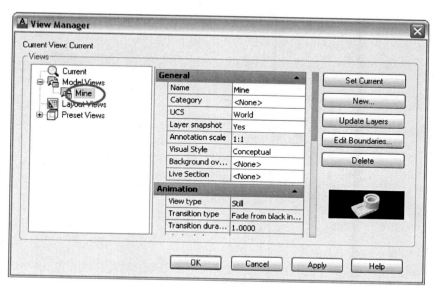

- Another place you will see a list of the saved views is the Views panel; see the following (you should save the file first):

9.2 3D AND VIEWPORT CREATION

- In this section we will discuss the special features of 3D in viewport creation. To issue this command, type at the Command window **+vports**, then press [Enter]. Press [Enter] again to accept the default value for the next prompt, and you will see the following dialog box:

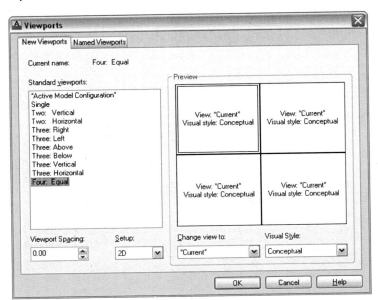

- Changing the **Setup** allows you to choose either 2D or 3D. When you choose 3D the arrangement under **Preview** will change to something like the following:

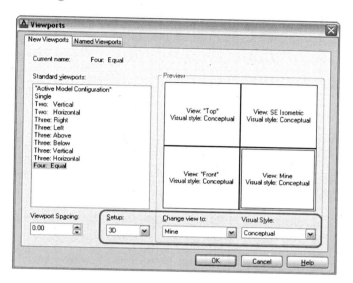

- Notice that in the illustration above, AutoCAD assigned a view for each viewport. You can change the default view, and you can also assign a saved view for the viewports using the **Change view to** list. You can also change the Visual Style. You will get something like the following:

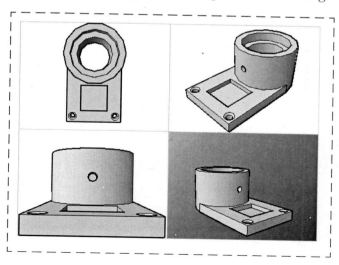

- You can assign a shade mode to one or all of the viewports. Select the desired viewport, then right-click, select **Shade plot**, and then select one of the modes as shown below:

- Notice the last group of the shade plots. They are rendering options, so you can plot a rendered image if you want, but you will only see the effects of this action using the **Print Preview** command.

VIEWS AND VIEWPORTS IN 3D

Practice 9-1

1. Start AutoCAD 2013.
2. Open **Practice 9-1.dwg**.
3. Change the visual style to Hidden.
4. Using the different viewing commands, try to get the following view or something close to it:

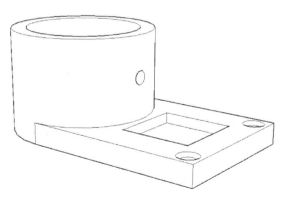

5. Save this view as "My First View."
6. Change the background to Gradient, and set the angle to 45°.
7. Go to Layout1.
8. Start the **+vports** command.
9. Select Four: Equal, then change the setup to 3D. Select the lower-right viewport and change the view to My First View, then make sure the visual style for the other three viewports is Conceptual.
10. Insert the four viewports in the space available in Layout1.
11. Select the upper-right viewport and change the shade plat to Presentation.
12. Using Plot Preview check the output; you should have something like the following:

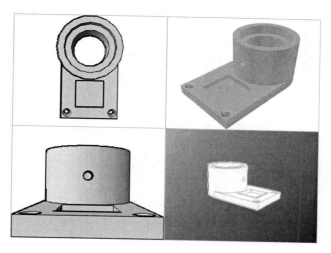

13. Save and close the file.

9.3 MODEL DOCUMENTATION – INTRODUCTION

- Model Documentation will help you generate intelligent documentation for AutoCAD 3D models. Model Documentation will generate views, sections, and details, which will be instantly updated when the model changes. You will find all the desired commands in the **Layout** tab: Create View, Modify View, Update, and Style and Standards.

- The procedure is very simple:
  - Create a Base view
  - Create Projected views
  - Create Sections and Details
  - Edit and update the views
  - Control Styles and Standards

9.4 MODEL DOCUMENTATION – CREATING A BASE VIEW

- Creating a Base view should always be your first step. If you have more than one 3D model, then AutoCAD allows you to select the desired objects to be included in the creating process. You can begin in Model Space, or you can go to the desired layout to create all the desired views. To start this command, go to the **Layout** tab, locate the **Create View** panel, and select the **Base** button:

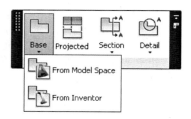

- You have two choices here:
  - Either from the current Model Space.
  - Or, from Inventor. This option allows you to import an Inventor file (*.iam, *.ipt, or *.ipn).
- Let's assume that you want to work with 3D objects from AutoCAD that were created in Model Space. In this case, we have two choices:
  - Either you will start the command in Model Space, and AutoCAD will ask you either to select the desired objects to include in the view/section/detail, or you will select the entire model. Because you are in Model Space, AutoCAD will ask to specify the name of the layout you will insert the different views/sections/details in. You will see the following prompts:

```
Select objects or [Entire model] <Entire model>: Enter new
or existing layout name to make current or [?] <Layout1>:
layout1
```

- Or, you will start the command in one of the layouts, and AutoCAD will ask you to specify the location of the Base view, or a group of other options. The most important option at this stage is the **Select** option. By default, AutoCAD will select all objects in the Model Space, but once you begin the Select option, you will be taken to Model Space to remove the undesired objects, and when done, AutoCAD will take you back to the layout you were at. You will see the following prompts:

```
Type = Base only Hidden Lines = Visible and hidden lines Scale
= 1/8" = 1'-0"
Specify location of base view or [Type/sElect/Orientation/
Hidden lines/Scale/Visibility] <Type>:
```

- In the meantime, a new context tab called **Drawing View Creation** will appear, and it looks like the following:

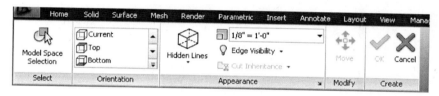

- At the left you can see the Model Space Selection button, which is identical to the Select option we discussed above.
- The default view will appear, but you can change it to your desired view, which will act as your Base view, using two methods. The first method is using the **Orientation** panel in the context tab, which allows you to select the desired view. The second method is to use the **Orientation** option in the prompts above.
- By default, using the Base command will insert the Base view and the projected views as well. But using the Type option you can select whether to insert Base or Base and Projected.
- Using the **Appearance** panel in the context tab allows you choose the Hidden Lines settings; you will choose one of four: Visible lines, Visible and hidden lines, Shaded with visible lines, and Shaded with visible and hidden lines. Using the same panel, you can set the scale and the edge visibility.

- If you only inserted the Base view, you will see something like the following:

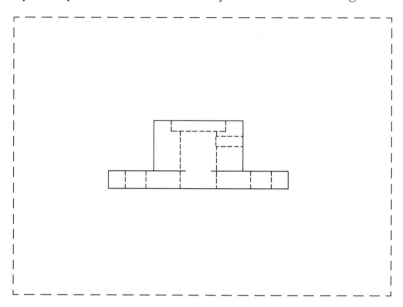

9.5 MODEL DOCUMENTATION – CREATING A PROJECTED VIEW

- If you create a Base View, without creating Projected Views as we did in the previous step, you still can create them at your convenience. Go to the **Layout** tab, locate the **Create View** panel, and select the **Projected** button:

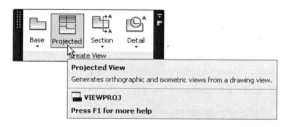

- The following prompts will be shown:

```
Select parent view:
Specify location of projected view or <eXit>:
Specify location of projected view or [Undo/eXit] <eXit>:
```

- Select the parent (Base) view and start creating projected views from it. You can create up to eight views; four are orthographic and four are isometric. You will see something like the following:

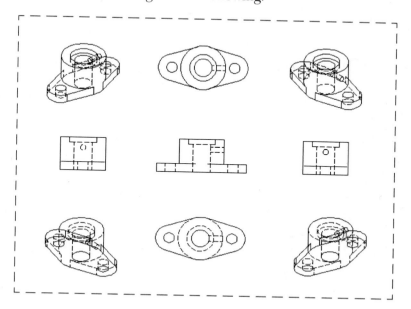

9.6 MODEL DOCUMENTATION – CREATING A SECTION VIEW

- After you insert the Base view, AutoCAD can produce section views. To use the section commands, go to the **Layout** tab and locate the **Create View** panel; click the list of **Section**, and you will see something like the following:

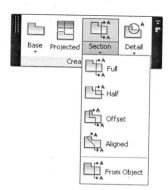

- There are five methods to create a section from a Base view:
  - Full section
  - Half section
  - Offset section
  - Aligned section
  - Section from object
- Using Full section you will see the following prompts:

```
Specify start point:
Specify end point or [Undo]:
Specify location of section view or:
```

- You should specify the start and end point of the section line and then specify the location of the section. You will get something like the following:

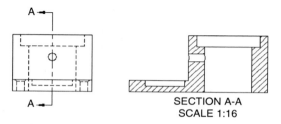

SECTION A-A
SCALE 1:16

- A half section is a view of an object showing one-half of the view. You will see the following prompts:

```
Specify start point:
Specify next point or [Undo]:
Specify end point or [Undo]:
Specify location of section view or:
```

- You will specify three points, then the location of the section. You will see something like the following:

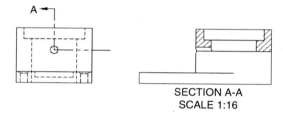

SECTION A-A
SCALE 1:16

An Offset section is a section view resulting from a line that does not follow a straight line to show features that do not reside on a straight line. You will see the following prompts:

```
Specify start point:
Specify next point or [Undo]:
Specify next point or [Undo]:
Specify next point or [Undo]:
Specify next point or [Undo/Done] <Done>:
```

- You will see something like the following:

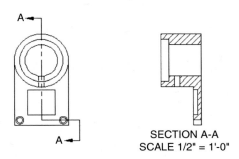

SECTION A-A
SCALE 1/2" = 1'-0"

- Aligned section will create a section line with an angle to include features of the Base view. You will see the following prompts:

```
Specify start point:
Specify next point or [Undo]:
Specify next point or [Undo]:
Specify next point or [Undo/Done] <Done>:
```

- You will see something like the following:

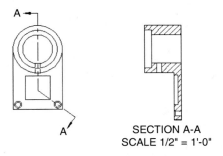

SECTION A-A
SCALE 1/2" = 1'-0"

- Finally, you can convert an existing object to act as a section line. You will see the following prompts:

```
Select objects or [Done] <Done>:
Select objects or [Done] <Done>:
Specify location of section view or:
```

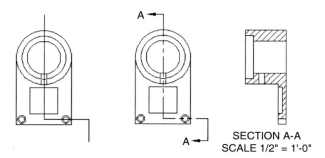

SECTION A-A
SCALE 1/2" = 1'-0"

- ■ When you use any of the following methods, a context tab called **Section View Creation** will appear, and it looks like the following:

- ■ You can control the following:
  - • Appearance (Hidden Lines, Scale, Edge Visibility)
  - • Annotation (Identifier, show or hide label)
  - • Hatch (show or hide hatch)

9.7 MODEL DOCUMENTATION – SECTION VIEW STYLE

- ■ All the things that appear when the section is created are controlled by the **Section View Style**. To activate this command, go to the **Layout** tab, find the **Styles and Standards** panel, and then click the **Section View Style** button:

- The following dialog box will appear:

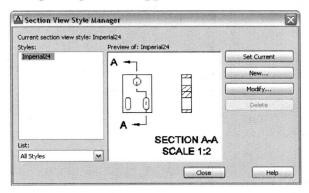

- You can create a new style or modify an existing one (there is a premade style called Imperial 24). If you click the **New** button, you will see the following dialog box:

- Input the name of the new style, click the **Continue** button, and you will see the following:

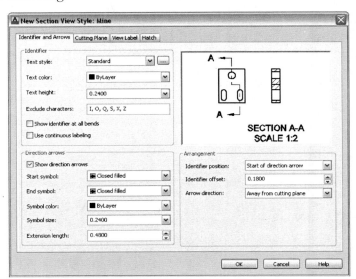

- There are four tabs in this style dialog box:
  - Identifier and Arrows
  - Cutting Plane
  - View Label
  - Hatch
- Each will be discussed as follows.

9.7.1 Identifier and Arrows Tab

- You will see the following:

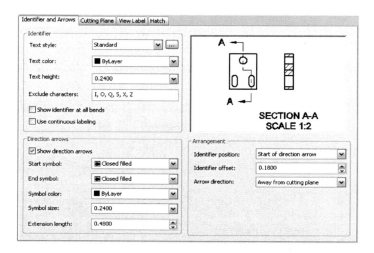

- In this tab, you can control:
  - Text style, Text color, and Text height
  - Characters to be excluded from the identifier naming (the default is to exclude I, O, Q, S, X, Z)
  - Whether to show the identifier at all bends of the section line
  - Whether to use continuous labeling
  - Whether to show the arrows direction
  - Start and end symbols
  - Symbol color and size
  - Extension length of the line before the arrow
  - Identifier position; there are five to choose from:

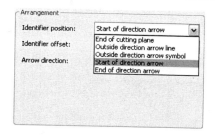

- Identifier offset (the distance to be left before inserting the identifier)
- Arrow direction; you have two choices:

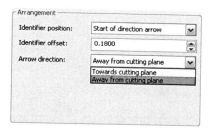

9.7.2 Cutting Plane Tab

- You will see the following:

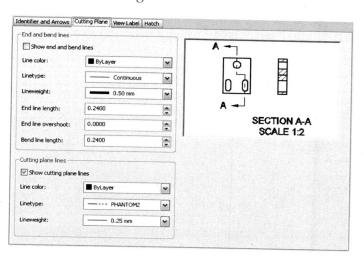

- In this tab, you can control:
  - Whether to show end and bend lines
  - Line color, Linetype, Lineweight

- End line length, which sets the length of the end line segment
- End line overshoot, which sets the length of the extension beyond the direction arrows
- Bend line length, which sets the length of the bend line segment on either side of the bend vertices
- Whether to show cutting plane lines or not, and then the color, linetype, and lineweight

9.7.3 View Label Tab

- You will see the following:

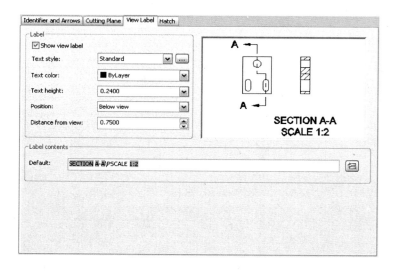

- In this tab, you can control:
  - Whether to show or hide the view label
  - Text style
  - Text color and Text height
  - Position of the label, either above or below
  - Distance of the label from the view
  - Default content

9.7.4 Hatch Tab

- You will see the following:

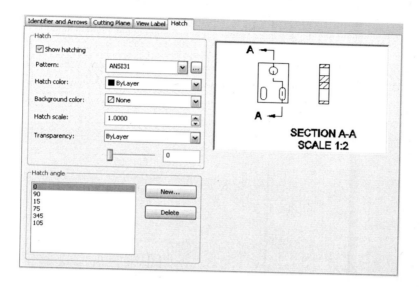

- In this tab, you can control:
  - Whether to show or hide hatch
  - Hatch pattern, hatch color, hatch background color, hatch scale, and hatch transparency (either leave it as ByLayer or specify a value)
  - Hatch angle; select one of the existing values or define your own using the New button

9.8 MODEL DOCUMENTATION – CREATING A DETAIL VIEW

- AutoCAD can produce detail views to magnify small details. To use the detail commands, go to the **Layout** tab, locate the **Create View** panel, click the list of **Detail**, and you will see something like the following:

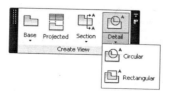

- You have two choices:
  - Circular
  - Rectangular
- In both types you will see the following prompt:

```
Specify center point or [Hidden lines/Scale/Visibility/
Boundary/model Edge/Annotation] <Boundary>
```

- Here is an example of the circular type:

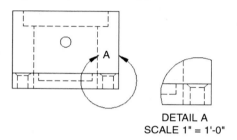

DETAIL A
SCALE 1" = 1'-0"

- And another example for the rectangular type:

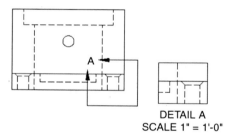

DETAIL A
SCALE 1" = 1'-0"

- When you use any of the discussed methods, a context tab called **Detail View Creation** will appear, and it looks like the following:

- You can control the following:
  - Appearance (Hidden Lines, Scale, Edge Visibility)
  - Boundary (the selected type is the type you started with)
  - Model edge (you have four choices to choose from: Smooth, Smooth with border, Smooth with connection line, and Jagged)
  - Annotation (Identifier, show or hide label)

9.9 MODEL DOCUMENTATION – DETAIL VIEW STYLE

- Everything that appears when the details are created is controlled by **Detail View Style**. To activate this command, go to the **Layout** tab, locate the **Styles and Standards** panel, and then click **Detail View Style** button:

- The following dialog box will appear:

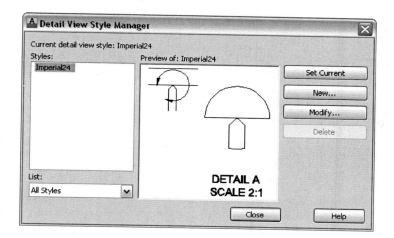

- You can create a new style or modify an existing one (there is a premade style called Imperial 24). If you click the **New** button, you will see the following dialog box:

- Input the name of the new style, and click the **Continue** button to start editing the style. You will see three tabs:
  - Identifier
  - Detail Boundary
  - View Label

9.9.1 Identifier Tab

- You will see something like the following:

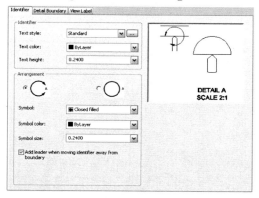

- In this tab, you can control:
  - Text style the identifier will be written with, the text color, and the text height
  - One of the two available arrangements
  - Symbol, symbol color, and symbol size, and whether to add a leader when moving the identifier away from the boundary

9.9.2 Detail Boundary Tab

- You will see something like the following:

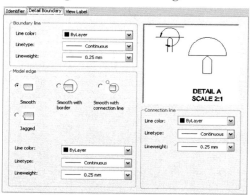

- In this tab, you can control:
  - The boundary line's color, linetype, and lineweight
  - The model edge (select one of the four available methods), then specify line's color, linetype, and lineweight)
  - If there will be a connection line, the color, linetype, and lineweight

9.9.3 View Label Tab

- You will see something like the following:

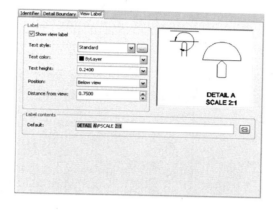

- In this tab, you can control:
  - Whether to show or hide the view label
  - Text style
  - Text color and Text height
  - Position of the label, either above or below
  - Distance of the label from the view
  - Default content

9.10 MODEL DOCUMENTATION – EDITING VIEW

- AutoCAD allows you to edit the view after inserting it. A special context tab will appear depending on the type of the view. To activate the **Edit View** command, go to the **Layout tab**, locate the **Modify View** panel, and then select the **Edit View** button:

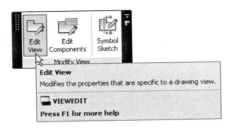

- You will see the following prompt:

```
Select view:
```

- Select the desired view (base, section, or detail). For Base view you will see the **Drawing View Editor** context tab:

- All of the buttons have already been discussed. For section view you will see the **Section View Editor** context tab:

- All of these buttons have also been discussed. For detail view you will see the **Detail View Editor** context tab:

- All of these buttons have also been discussed.
- Note the following:
  - The same effect will take place if you double-click the desired view.
  - Projected view will show the same Drawing View Editor context tab, except the Selection button will be off.

- If you are in Base view, the Quick Properties window will look like the following (you can change both the rotation angle and scale):

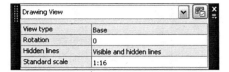

- For Projected view, you will see the following (you can break the alignment and change the angle):

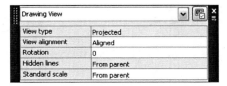

- For Section view, you will see the following:

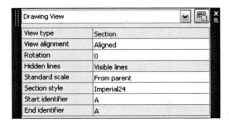

- For Detail view, you will see the following:

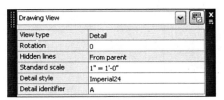

9.11 MODEL DOCUMENTATION – EDITING COMPONENTS

- AutoCAD allows you to edit any section participation. To activate the **Edit Components** command, go to the **Layout** tab, locate the **Modify View** panel, and then select the **Edit Components** button:

- You will see the following prompt:

```
Select component:
Select section participation [None/Section/sLice] <None>: S
```

- There are three options to choose from:
  - None
  - Section
  - Slice
- The following three illustrations show you the difference:

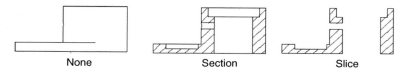

 ■ Note that the Symbol Sketch command is used to constrain the Section and Detail view to the Drawing view. But in order to do that you have to know the constraint types of AutoCAD (i.e., Geometry and Dimensional).

9.12 MODEL DOCUMENTATION – UPDATING VIEWS

- After you insert base, projected, section, and detail views, you may need to change the original drawing. But what will happen to the views? AutoCAD has two methods to update the views: Automatic Update (by default it is on) and Manual update. You can opt to turn Automatic Update off and manually update the different views. To use these two features, go to the **Layout** tab, locate the **Update** panel, and use either the **Auto Update** button (which is on by default) or use the **Update View** list, which has **Update View** and **Update All Views** buttons:

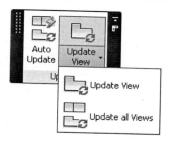

- If the Automatic method is used, then whenever you make a change to the original model, the views will change, and you will see a prompt like the following:

```
2 view(s) updated successfully.
```

- The number given depends on the number of views you have in the current layout(s). If you use the manual method, the following will take place:
  - You will see a red frame around the views that should be updated; something like the following:

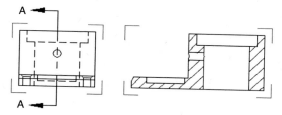

  - A bubble will appear at the lower-right of the screen; something like the following. Select Update all drawing views on this layout in the bubble in order to update them, or click the button on the Update panel.

  - Or select view(s), right-click, then select the **Update View** option:

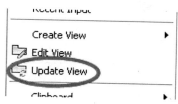

- Finally, if you go to the right of the status bar, you will see something like the following:

- If you right-click this button, you will see the following menu, which allows you to update a single view or all views:

9.13 MODEL DOCUMENTATION – DRAFTING STANDARDS

- This dialog box controls the output of the Drawing Views commands. To issue this command, go to the **Layout** tab, locate the **Styles and Standards** panel, and then select the **Drafting Standards** button:

- You will see the following dialog box:

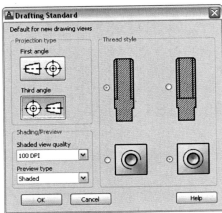

- Under **Projection type**, you will define where the projected views are placed. You will have two choices:
  - First angle means top views are placed below the Base view.
  - Third angle means top views are placed above the Base view.
- Under **Shading/Previewing**, set the resolution and type of the view.
- Finally, this command can be used to important Inventor files. If this is the case, specify the **Thread style.**

9.14 MODEL DOCUMENTATION – ANNOTATION MONITOR

- After you insert base, projected, section, or detail views, you can put dimensions on these using normal dimension commands. These dimensions will be fully associated with the view, and if it is moved or rotated, they will move or rotate as well even if they were not selected. But what if one or more of the features the dimensions were based on is changed or removed? AutoCAD implemented a tool for this called **Annotation Monitor**, which is an icon that appears in the status bar:

- If this button is on, it will monitor any dimension in the views; if for any reason the dimension becomes disassociated with the model, the Annotation Monitor will be shown at the right of the status bar in red. Alert badges will be displayed on the dimensions that were disassociated; something like the following:

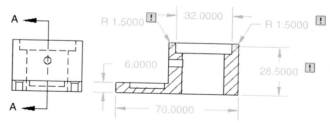

- And a bubble will appear, alerting you; something like the following:

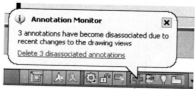

- You can use the last line to delete the disassociated annotations, or you can do this manually. If you click on the alert badge you will see the following menu:

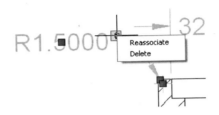

- Selecting the **Delete** option will delete this dimension, but if you select the **Reassociate** option, you will be allowed to select new points (or objects or radius and diameter).
- Also note the following:
  - You can manipulate the section line using the normal grips.
  - AutoCAD will create several layers once you add a Base view, projected view(s), section view(s), and detail view(s), MD_Annotation, MD_Hatching, MD_Hidden, and MD_Visible, and will put each part in its respective layer, except your dimensions. You should make MD_Annotation the current layer.

MODEL DOCUMENTATION

Practice 9-2

1. Start AutoCAD 2013.
2. Open **Practice 9-2.dwg**.
3. You will see two parts.
4. Go to Part (1) layout.
5. Go to Layout tab, locate the Create View panel, select the Base button, and then select From Model Space.
6. You will see the two parts selected; right-click and choose the Select option. This will take you to the Model Space; select the remove option and select the part at the left to be removed, then press [Enter].
7. Using the context tab, make sure that Front and Hidden Lines are selected, then set the scale to ½" = 1'.
8. Insert the view at the middle center of the sheet and one projected view to the left, then press [Enter] to end the command.

9. Start Section View Style and modify the existing style using the following:
    a. Text height for identifier = 0.14
    b. Under Direction arrows start symbol and end symbol should be closed blank
    c. Symbol size = 0.12
    d. Text height for view label = 0.14 and distance from view = 0.4
    e. Under Hatch scale = 0.75
10. Using Full Section, add a section to the right of the Base view cutting the middle of the Base view (use OTRACK and the circle at the middle).
11. Start Detail View Style and modify the existing style using the following:
    a. Text height for identifier = 0.14
    b. Arrangement = Full circle
    c. Under Detail Boundary select Smooth with connection line
    d. Text height for view label = 0.14 and distance from view = 0.4
12. Add a circular detail for the screw hole at the right, below the Base view, using scale 1" = 1' (if the text goes beyond the dashed line, you can move it up).
13. Turn off the Auto Update button.
14. Go to Model Space.
15. Click the Home button of the ViewCube.
16. Zoom in to the top of the shape at the right and fillet the outside edge of the top using radius = 1.5.
17. Go to Part (1) layout.
18. Select to update all views.
19. Zoom in to the section at the right.
20. Make MD_Annotation the current layer.
21. Using the normal dimension add the following dimensions:

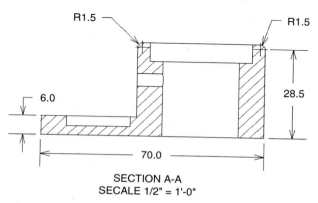

SECTION A-A
SECALE 1/2" = 1'-0"

22. Go back to Model Space and delete the fillet you added in step 16.

23. Go to Part (1) layout and update all views.
24. Notice the alert badges in the drawing and the bubble that appeared in the Annotation Monitor. Close the bubble without using the link.
25. Manually delete the two radius dimensions.
26. Reassociate the linear dimension at the right with the full length.
27. Go to Part (2) layout and create a similar layout for the part at the left using scale ¼" = 1', and select proper scales for the other views.
28. Save and close the file.

9.15 3D DWF CREATION AND VIEWING

- You have the ability to produce a 3D DWF file that can be viewed in Autodesk Design Review. A 3D DWF allows you to look at the 3D model from different angles, dismantle the 3D model, and also take sections. There are two methods to produce a 3D DWF:
  - Single sheet 3D DWF using the 3D DWF command.
  - Multiple sheet 3D DWF using the Batch Plot command.

9.15.1 3D DWF Command

- This command allows you to produce a single sheet 3D DWF file. To issue this command, go to the **Output** tab, locate the **Export to DWF/PDF** panel, and then select the **3D DWF** button:

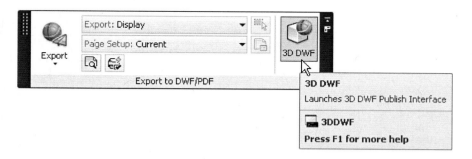

- The following dialog box will appear:

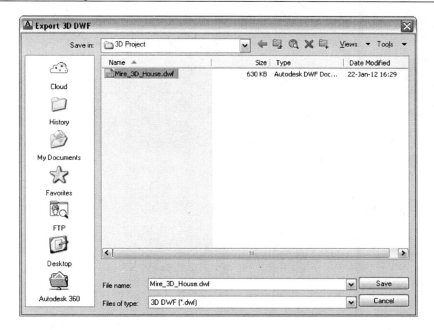

- When AutoCAD has finished producing the file, you will see the following message:

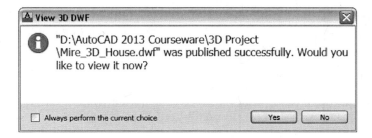

- If you select Yes, AutoCAD will open Autodesk Design Review to view the 3D file.

9.15.2 Batch Plot Command

- This command allows you to produce a multiple sheet 3D DWF file. To issue this command, go to the **Output** tab, locate the **Plot** panel, and select the **Batch Plot** button:

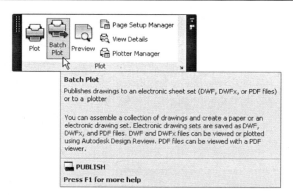

- The following dialog box will appear:

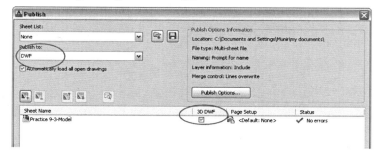

- Under **Publish to** select DWF or DWFx. For Model sheet, you should turn on the checkbox of 3D DWF. This is the only way to tell AutoCAD you want the 3D DWF. To control the 3D DWF, click the **Publish Options** button, and the following dialog box will appear:

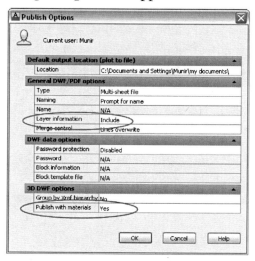

- Make sure the **Layer information** value is **Include**. At the bottom of the dialog box make sure the **Publish with materials** value is **Yes**.
- Finally, you should understand that not all DWG data will be transferred to a 3D DWF, including:
  - Animation
  - Some Fonts
  - Lights and shadow
- To see a full list consult the AutoCAD Help and manual.
- After completing the production of the 3D DWF file, start Autodesk Design Review to view it. You will see something like the following:

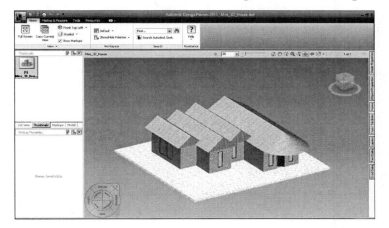

- At the left part of the screen you will see a small window called **Model**. This window contains all the layers in your original AutoCAD file. Clicking one of the layers will list the solids in it, and this is a good way to select solids, if you don't want to use the mouse to pick up objects.

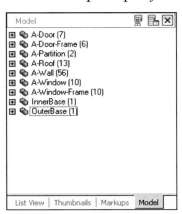

- Also notice the ViewCube at the top-right portion of the screen. Other viewing commands are shown at the top canvas. If you select the **Tools** tab, you will see two active panels, **3D Tools** and **Create Sheet**, which include the following buttons:
  - Move & Rotate
  - Sectioning buttons
  - Create a sheet in the DWF file from a snapshot

9.15.3 Move and Rotate Commands

- This command will help you take the model apart by moving and rotating the different parts of the model. To issue this command, go to the **Tools** tab, locate the **3D Tools** panel, and then select the **Move & Rotate** button:

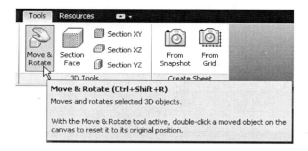

- Once this command is working, the cursor will change to the shape of a hand, and whenever the hand hovers over any 3D object, it will highlight it. Once highlighted, this will enable you to select the object with a simple click. The desired object will change to something like the UCS icon. Using this icon you can move in the directions of X, Y, Z, XY, YZ, and ZX, and rotate around X, Y, and Z. See the following illustrations:
  - Moving along the Z-axis:

- Moving in the Y-Z plane:

- Rotating around the Z-Axis:

- To get things back to normal, do one of the following two things:
    - To return a single object to its original place, double-click on the object.
    - To return all objects to their original places, click the **Home** icon on the canvas.

9.15.4 Sectioning

- This command allows you to create a section of the 3D DWF file:
    - Parallel to an existing face
    - Along XY, XZ, and YZ
- To issue this command, go to the **Tools** tab, locate the **3D Tools** panel, and then select any of the buttons shown below:

- If you select the first option, you will see the mouse cursor change to a knife. Select the desired face to be your cutting face, and when you click it the section will be created. Then the mouse will change to a hand holding a cube, along with the UCS icon. Move the hand in the direction you want to cut your model. If you select the second option, the cutting plane will be placed at the middle of your model. You will then see the hand holding the cube along with the UCS icon to be moved to cut your model and get the desired section. See the following example, which shows the XY section from the top view:

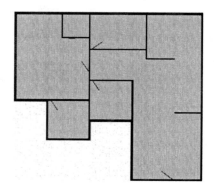

- Note the following:
  - You can use multiple sections at the same time.
  - Pressing [Esc] will retain the section plane in its current position.
  - Clicking **Home** from the canvas will return things to normal.

9.15.5 Taking a Snapshot

- After setting up your desired view, and/or section, you can take a snapshot and add it to the list of sheets in the current DWF file. To issue this command, go to the **Tools** tab, locate the **Create Sheet** panel, and then select the **From Snapshot** button:

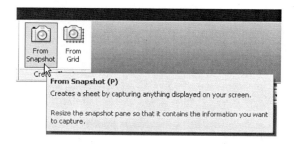

- You will see a red frame showing the current resolution of the snapshot. You can resize the frame using all the sides and all the corners. See the following example, which shows the initial size of the red rectangle and how to resize it:

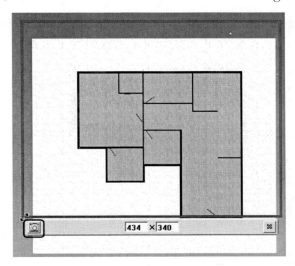

- If everything is OK, click the small button with the camera on it at the lower-left corner of the frame to create a new sheet.

PRODUCING AND VIEWING A 3D DWF FILE

 Practice 9-3

1. Start AutoCAD 2013.
2. Open **Practice 9-3.dwg**.
3. Using the 3D DWF button, create a new 3D DWF file and call it *3D Model.dwf*.
4. Open Autodesk Design Review and open 3D Model.dwf.
5. Using the different viewing commands, view the model from different angles.
6. Using the Move & Rotate command, dismantle the model to its components, then double-clicking and using the Home button, return things to normal.
7. Using the four section commands, section your model several times.
8. Using XZ section and top view, save the view as a new sheet in the current 3D DWF file.
9. Save and close the file.

NOTES:

CHAPTER REVIEW

1. In Model Documentation, my Base view should always be Top view.
   a. True
   b. False
2. In Autodesk Design Review you can create a section in 3D DWF file using 3 points.
   a. True
   b. False
3. Which one of the following statements is *not* true about views and viewports in 3D?
   a. You can assign a background for a view.
   b. You can assign render settings for a viewport.
   c. You can't save the visual style with a view.
   d. You can show a saved view in the viewport display.
4. There are two methods to create a 3D DWF file. Which one of the following statements is correct?
   a. Using the 3D DWF button from the Output tab and the Plot panel.
   b. Using the Batch Plot button from the Output tab and the Export to DWF/PDF panel.
   c. Using the 3D DWF button from the Home tab and the Export to DWF/ PDF panel.
   d. Using the Batch Plot button from the Output tab and the Plot panel.
5. The maximum number of views you can produce per sheet is _____ views.
6. You can insert a Base view, then create _____, _____, and _____ views from it.

CHAPTER REVIEW ANSWERS

1. b
3. c
5. Nine

10 CAMERAS AND LIGHTS

Chapter

In This Chapter
◇ Creating and controlling cameras
◇ Creating and editing lights

10.1 INTRODUCTION

- Chapter 10 and Chapter 11 are closely connected because if you master the content in both, you will be able to produce professionally rendered images. In this chapter we will discuss the first two steps of this process:
  - Creating, setting up, and editing camera(s) in a drawing.
  - Creating, setting up, and editing the light(s) in a drawing.
- Once you master these two steps, you will be ready for Chapter 11 where we will work with materials and produce rendered images.

10.2 HOW TO CREATE A CAMERA

- To create a camera in AutoCAD, you will first specify two positions in the XYZ space; one for the camera and one for the target. You will need to know the limits of your model in the XY plane first, then the best Z value for looking at your model. Using OSNAP is another way to specify the two positions. When the creation process is successful, the Field-of-View (FOV) will be created based on the camera lens length. See the illustration below:

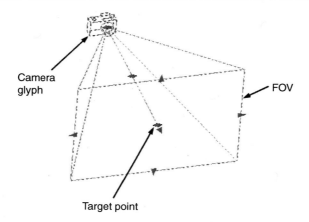

- To create a new camera, go to the **Render** tab, locate the **Camera** panel, and then select the **Create Camera** button:

- The following prompts will be shown:

```
Current camera settings: Height=0 Lens Length=50.0 mm
Specify camera location:
Specify target location:
Enter an option [?/Name/LOcation/Height/Target/LEns/Clipping/
View/eXit]<eXit>:
```

- The first line gives the current value for camera height and lens length. AutoCAD then asks you to specify the coordinates of both the camera and the target points. You can type the coordinates or specify points on the screen using OSNAP. Once you are done, you can end the command, or change one of the following parameters:
  - Name: to change the name of the camera from the default name Camera 1, etc.
  - Location: to change the location of the camera point.
  - Height: to change the camera height.
  - Target: to change the target position.

- Lens: to change the lens length of the camera. The greater the lens length, the smaller the FOV.
- Clipping: will be discussed in the next part of this chapter.
- View: to switch to the camera view.
- Exit: to end the command.

10.3 HOW TO CONTROL A CAMERA

- After the creation process ends, you will see a camera glyph in the drawing. If you click the camera glyph two things will happen:
  - The **Camera Preview** dialog box will appear, which shows the camera view. You can select the visual style to be used.
  - Grips will appear showing the camera and target points, along with the FOV.

10.3.1 Camera Preview Dialog Box

- Whenever you will click the camera glyph in the drawing, the **Camera Preview** dialog box will appear showing the current camera view. You have the ability to change the visual style in this dialog box. You can also specify whether to show this dialog box in the future. See the below illustration:

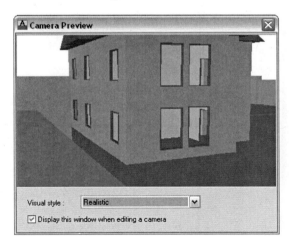

10.3.2 Grips

- Whenever you select a camera glyph, grips along with a pyramid of a rectangular base and an apex point will appear. The apex point is the camera point, and at the center of the rectangular base is the target point. The pyramid represents the FOV. The following illustrates the concept:

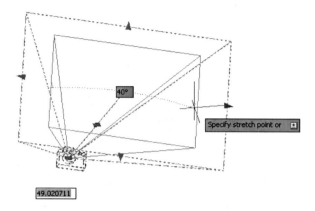

- If you select a camera and right-click and select the **Properties** option the **Properties** palette of the camera, just like the following, will appear:

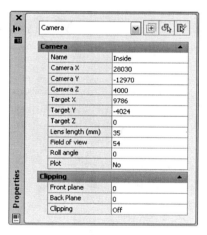

- As you can see, you can change all or any of the camera properties, beginning with the camera name and ending with enabling the front and back clipping planes.

10.3.3 Front and Back Clipping Planes

- The front and back clipping planes are imaginary clipping planes. When you change their current position they will hide some or all of the view. Objects between the camera position and the front clipping plane are not visible; objects between the target and the back clipping plane are not visible either.

- You can turn the planes on/off using the **Properties** palette. See the following:

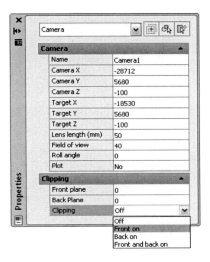

- Moving the front clipping plane away from the camera will obscure some of the model; but moving the back clipping plane closer to the camera will also obscure some of the model. It is best to change the view of your model to Top view in order to get the benefits of the clipping. Using this view allows you to identify both the front clipping plane and back clipping plane easily. The following two illustrations help explain the concept:

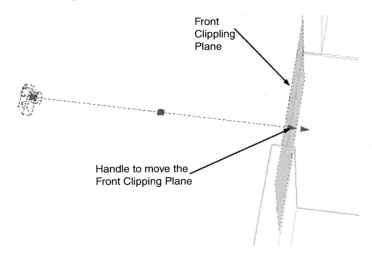

- By moving one of the two clipping planes, and using the Camera Preview dialog box, you will see how the clipping affects the scene.
- If the clipping plane is enabled its effects will always be shown. To remove the clipping effects, you should turn the clipping plane off.

10.3.4 Camera Display and Camera Tool Palette

- By default the camera glyph is shown, but you can hide it by going to the **Render** tab, locating the **Camera** panel, and selecting the **Show Camera** button. This button will turn the camera glyph on/off:

- AutoCAD comes with several predefined cameras ready for use. You can access these cameras using Tool Palettes. Search for a palette called **Cameras,** which includes three cameras:
  - Normal Camera (50 mm lens length)
  - Wide-angle Camera (35 mm lens length)
  - Extreme Wide-angle Camera (6 mm lens length)
- See the following illustration:

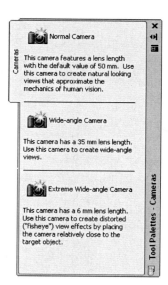

CREATING AND CONTROLLING CAMERAS

Practice 10-1

1. Start AutoCAD 2013.
2. Show the Camera tool palette.
3. Using the Wide-Angle Camera, create a camera using the line at the entrance, using the outside endpoint as the camera point and the other end as the target point. Name it "Entrance."
4. Thaw layer OuterBase.
5. Preview the camera using different visual styles.
6. Using the Extreme Wide-Angle Camera, create a camera using the column at the far right end to the top of the roof of the entrance:

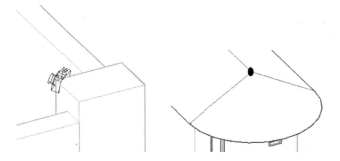

7. Preview the camera using the Conceptual visual style, and notice how the picture is distorted.
8. Create a third camera using your own set of points, choosing the desired camera type.
9. Save and close the file.

10.4 INTRODUCING LIGHTS

- AutoCAD files come with two default distant light sources following the viewpoint as it moves. AutoCAD will ask you to turn these two default lights off when you attempt to add a light.
- The following lights can be added:
  - Pointlight and Target Pointlight
  - Spotlight and Free Spotlight
  - Distant Light
  - Weblight and Free Weblight

- The first step before inserting any light is to set the lighting units. You can do this by going to the **Render** tab, locating the **Lights** panel, and then selecting the last pop-up list, as follows:

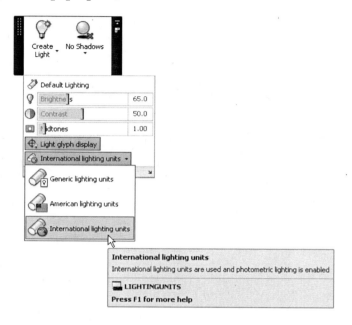

- The following light units are available:
  - Generic lighting units: lights will be normal light sources.
  - American lighting units: lighting units are foot candles and lights will be photometric.
  - International lighting units: lighting units are lux and lights will be photometric.
- Photometric lights:
  - Are real-world lighting and allow you to produce realistic rendered images.
  - Use light energy values, which allow you to define lights in a practical way.
  - Allow you to import *IES* files and use specific light data from manufacturers.

10.5 INSERTING POINTLIGHT

- In AutoCAD there are two types of pointlight:
  - Pointlight
  - Target Pointlight

- Pointlight has no target point; it emits light in all directions. Target Pointlight, however, emits light to a certain target point. To issue this command, go to the **Render** tab, locate the **Lights** panel, and then select the **Point** button:

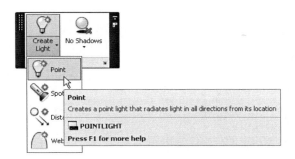

- The following prompts will be shown:

```
Specify source location <0,0,0>:
Enter an option to change [Name/Intensity factor/
Status/Photometry/shadoW/Attenuation/filterColor/eXit]
<eXit>:
```

- In order to locate the proper light location, identify the proper X,Y,Z point to specify the best light location. You will be able to change all or any of the following parameters.

10.5.1 Name

- You can name the light. You will see the following prompt:

```
Enter light name <Pointlight5>:
```

10.5.2 Intensity Factor

- You can change the value of the light intensity (brightness) factor. You will see the following prompt:

```
Enter intensity (0.00 - max float) <1.0000>:
```

10.5.3 Status

- You can turn the light ON or OFF. You will see the following prompt:

```
Enter status [oN/oFf] <On>:
```

10.5.4 Photometry

- You can change the Intensity and/or Color of the light. You will see the following prompt:

```
Enter a photometric option to change [Intensity/Color/
eXit] <I>:
```

- If you select Intensity, the following prompt will be shown:

```
Enter intensity (Cd) or enter an option [Flux/
Illuminance] <1500.0000>:
```

- Here, you can change the following:
  - Intensity: the power produced by a light source in a particular direction.
  - Flux: the apparent power per unit of solid angle.
  - Illuminance: the total luminous flux incident on a surface, per unit area.
- If the Flux option is selected, the following prompt will be shown:

```
Enter Flux (Lm) <18849.5559>:
```

- If the Illuminance option is selected, the following prompt will be shown:

```
Enter Illuminance (Lx) or enter an option [Distance]
```

- If the Color option is selected, the following prompt will be shown:

```
Enter color name or enter an option [?/Kelvin] <D65>:
```

- Here, color has no connection to the normal color of the light; the available colors are:

```
"D65" "Fluorescent" "Coolwhite" "Whitefluorescent"
"daylightfluorescenT" "Incandescent" "Xenon"
"Halogen" "Quartz" "Metalhalide" "mErcury"
"Phosphormercury" "highpressureSodium"
"Lowpressuresodium"
```

10.5.5 Shadow

- This option allows you to turn shadow on or off and specify the shadow type. You will see the following prompt:

```
Enter [Off/Sharp/soFtmapped/softsAmpled] <Sharp>:
```

- You can select one of three available shadow types:
  - Sharp: which is the lowest level of shadow.
  - Soft Mapped: which will produce realistic shadow with soft edges.
  - Soft Sampled: which is the best shadow type.

10.5.6 Attenuation

- This option allows you to specify the method of diminishing light over the distance. You will see the following prompts:

```
Enter an option to change [attenuation Type/Use limits/attenuation
start Limit/attenuation End limit/eXit] <eXit>:
```

- Type **T** to change the attenuation type, and the following prompt will be shown:

```
Enter attenuation type [None/Inverse linear/inverse Squared]
<None>:
```

- You can choose one of the three available types:
  - None: light will never diminish over distance.
  - Inverse Linear: light intensity will diminish using the following formula: $1/x$. x here is the distance.
  - Inverse Squared: light intensity will diminish using the following formula: $1/x^2$. x here is the distance.
- The final step is to specify whether to use limits for attenuation. Accordingly, specify the start and end limits for the light.

10.5.7 Filter Color

- This option allows you to set the light's color. You have three options to choose from:
  - True color (RGB); input a value for each color (0–255)
  - Index color (from 1–255)
  - HLS (Hue, Luminance, and Saturation)

10.5.8 Exit

- This option will end the command.

10.5.9 Target Pointlight

- Target pointlight is identical to pointlight with one exception; with this command, you will specify a target point. To issue this command, you should type **light** at the Command window, and accordingly the following prompts will be shown:

```
Specify source location <0,0,0>:
Specify target location <0,0,-10>:
Enter an option to change [Name/Intensity factor/Status/
Photometry/shadoW/Attenuation/filterColor/eXit] <eXit>:
```

10.6 INSERTING SPOTLIGHT

- AutoCAD offers two types of spotlights:
  - Spotlight: which uses a source and target point.
  - Free Spotlight: which uses a source point without a target point.
- To issue this command, go to the **Render** tab, locate the **Lights** panel, and then select the **Spot** button:

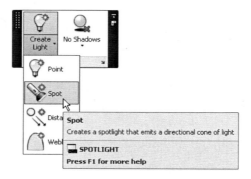

- You will be asked to specify two cones:
  - Hotspot cone
  - Falloff cone
- Light intensity will be the same in the Hotspot cone but will diminish starting from the Hotspot cone until the perimeter of the Falloff cone.
- The following illustrates the concept:

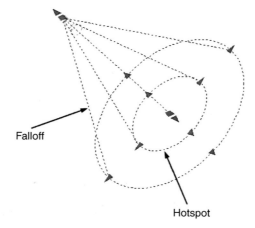

- The following prompts will be shown:

```
Specify source location <0,0,0>:
Specify target location <0,0,-10>:
Enter an option to change [Name/Intensity factor/Status/
Photometry/Hotspot/Falloff/shadoW/Attenuation/filterColor/eXit]
```

- AutoCAD asks you to specify a source position in XYZ and a target position. The rest of the options were all discussed with the Pointlight command, except the following two options.

10.6.1 Hotspot Option

- This option allows you to specify the angle of the Hotspot cone. The following prompt will be shown:

```
Enter hotspot angle (0.00-160.00) <45.0000>:
```

- As the prompt shows, you should input a value between 0 and 160°.

10.6.2 Falloff Option

- This option allows you to specify the angle of the Falloff cone. The following prompt will be shown:

```
Enter falloff angle (0.00-160.00) <50.0000>:
```

- As the prompt shows you should input a value between 0 and 160°.
- In order to reach Free Spotlight, type **light** at the Command window, and the following prompts will be shown:

```
Specify source location <0,0,0>:
Enter an option to change [Name/Intensity factor/Status/
Photometry/Hotspot/Falloff/shadoW/Attenuation/filterColor/eXit]
```

10.7 INSERTING DISTANT LIGHT

- Distant light is like the other sources of light and has a source position and a target position. Distant light will release light in a parallel fashion using the line between the source and target points. Every object in the path of the light will be affected. In order to issue this command, go to the **Render** tab, locate the **Light** panel, and then select the **Distant** button:

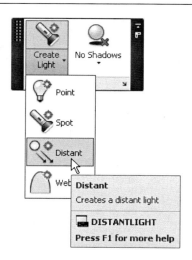

- The following prompts will be shown:

```
Specify light direction FROM <0,0,0> or [Vector]:
Specify light direction TO <1,1,1>:
Enter an option to change [Name/Intensity factor/Status/
Photometry/shadoW/filterColor/eXit] <eXit>:
```

- AutoCAD asks you to specify the position of the source and the target. The rest of the options were already discussed.

10.8 WORKING WITH SUNLIGHT

- AutoCAD gives you the option to see the effect of sunlight on your 3D model, in order to produce realistic rendering. To set your geographic location, go to the **Render** tab, locate the **Sun & Location** panel, and then select the **Set Location** button:

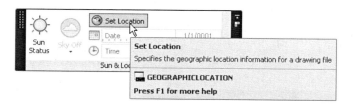

- The following dialog box will appear:

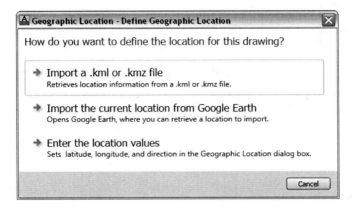

- This dialog box gives you three options to choose from:
  - To import *.kml or *.kmz files to retrieve the location information
  - To import the current location from Google Earth
  - To enter the location values manually
- Either import one of the two files, or enter your location manually. If you choose to enter your information, the following dialog box will appear:

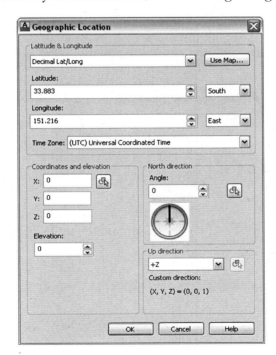

- If you know the latitude and longitude of your city, you can enter those coordinates; otherwise, use the **Use Map** button, and the following dialog box will appear:

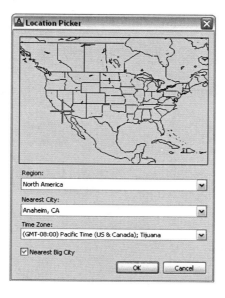

- You will specify Region, Nearest city, and Time Zone, and AutoCAD will calculate the position of the sun in relation to your city. Next, you must change the Sun Status to ON by going to the **Render** tab, locating the **Sun & Location** panel, and then selecting the **Sun Status** button:

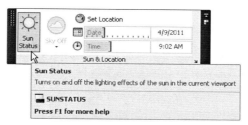

- To see the effect of the sun for different months and for different hours of the day, you should control the **Date** and **Time** using the same panel, and dragging the two sliders, as shown below:

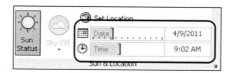

- You can also add the **Sky Effect** if it is an outside scene. You can select one of the following three:

- You can also control the **Sun Properties** by using the same panel and clicking the small arrow at the right:

- The sun palette will appear for editing:

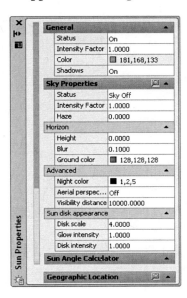

10.9 INSERTING WEBLIGHT

- All previous light types discussed are designed to produce light in a consistent fashion, but Weblight is designed to produce light in a non-uniform fashion. Using Weblights allows AutoCAD to produce rendered images that look and feel like the real world. To better understand this idea, imagine a sphere with pointlight distributed in a certain way (not necessarily covering the whole surface of the sphere). This is Weblight, and the distribution of the light will be produced using a file from the manufacturer. AutoCAD provides two types of Weblight:
  - Weblight: has a source and target points.
  - Free Weblight: doesn't have a target point.
- In order to issue this command, go to the **Render** tab, locate the **Light** panel, and then select the **Weblight** button:

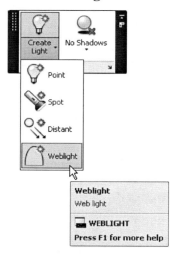

- The following prompts will be shown:

```
Specify source location <0,0,0>:
Specify target location <0,0,-10>:
Enter an option to change [Name/Intensity factor/Status/
Photometry/weB/shadoW/filterColor/eXit] <eXit>:
```

- All of these prompts have been discussed except the following option.

10.9.1 Web Option

- Choosing this option will show the following prompt:

```
Enter a Web option to change [File/X/Y/Z/Exit] <Exit>:
```

- Either type in the coordinates of the lights around the sphere (this is likely a very lengthy and tedious job!) or load the file that has the coordinates, which you can get from the manufacturer. Use the **Properties** palette to load the desired file.
- The following palette will be shown:

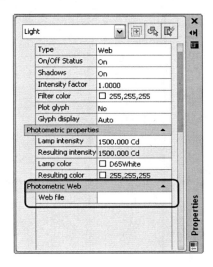

- Under **Photometric Web** click the three small dots to load the file (*.IES). AutoCAD comes with eight Web files. See the following illustration:

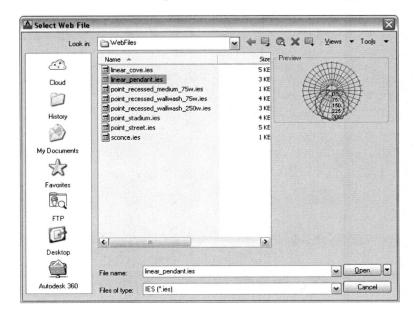

10.10 EDITING LIGHTS

- Show the Properties palette for both pointlight and spotlight to edit their properties:

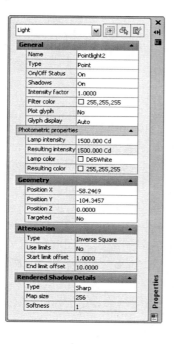 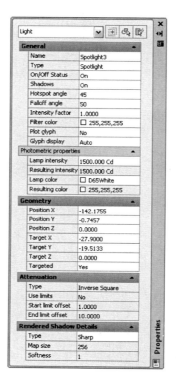

- Another way to edit spotlight is by using grips. You can edit the Hotspot cone, Falloff cone, and the Target point. The following illustrates the concept:

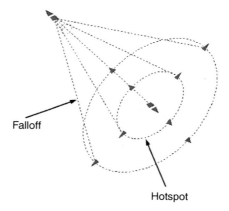

10.11 WORKING WITH PREMADE LIGHTS

- By default AutoCAD offers two palette groups:
  - Generic Lights
  - Photometric Lights
- The Photometric Lights group includes four palettes:
  - Fluorescent
  - High Intensity Discharge
  - Incandescent
  - Low Pressure Sodium
- See the following:

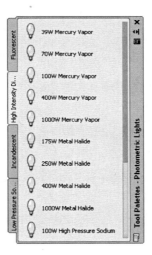

10.12 WHAT IS THE "LIGHTS IN MODEL" FUNCTION?

- This function allows you to identify the lights defined in the current drawing. To issue this command, go to the **Render** tab, locate the **Lights** panel, and then select the arrow at the right:

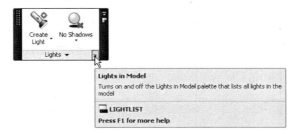

- The following palette will come up:

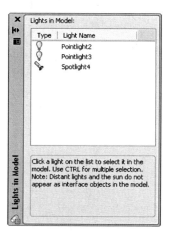

- Right-click any of the existing lights in the palette and a short-cut menu will appear like the following:

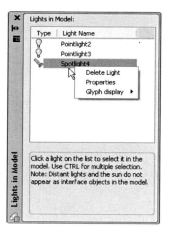

- Here, you can:
    - Delete a light
    - Show the Light Properties palette
    - Change the light glyph to Auto, On, Off
- NOTE ■ You will not be able to see the effects of the light on the model using these visual styles:
    - 2D Wireframe
    - Conceptual

- Hidden
- Sketchy
- Wireframe
- X-Ray

CREATING AND CONTROLLING LIGHTS

Practice 10-2

1. Start AutoCAD 2013.
2. Open **Practice 10-2.dwg**.
3. Notice that because we are in the Conceptual visual style we will not be able to see the effects of the light. So, change the visual style to Realistic.
4. Since there are no lights defined in this file, and the sunlight is not turned on, the generic two distant source lights are working. Turn on the sunlight.
5. The scene becomes a night scene with no lights. Set your location to Los Angeles.
6. Using the Light panel, turn on the Ground Shadows.
7. Move the Time and Date slider to see the effects of the light over the model.
8. Turn the Sun off.
9. Freeze layer Lamp_Cover.
10. Insert a pointlight at the top of the cylinder, without changing any of the default values.
11. Using the Photometric Light palette group, select the Fluorescent tab.
12. Select the 59W 8ft Lamp and add it at the middle of the top portion of the wall as shown below:

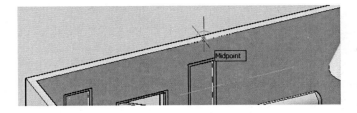

13. You need to rotate it by 90° to be parallel to the wall.
14. Move it to the inside of the room by distance 1600.
15. Then make two copies of it seperated by distance 1600.
16. Save and close the file.

NOTES:

CHAPTER REVIEW

1. FOV means Field of View.
   a. True
   b. False
2. Which one of the following statements is incorrect?
   a. There are two types of Spotlight.
   b. You can't see the effects of light in the Conceptual visual style.
   c. There is one type of Distant light.
   d. You can see the effects of light in the Conceptual visual style.
3. Which one of the following statements is *not* true?
   a. There are two cones for Spotlight.
   b. The greater the lens length, the smaller the FOV.
   c. FOV is a pyramid and the camera point is the apex point.
   d. You can't use Google Earth to set up your location.
4. Spotlight and No-Target spotlights are the two types AutoCAD default lights.
   a. True
   b. False
5. The file extension for Weblight files is _____.
6. The Extreme Wide-angle camera uses a _____ mm lens length.

CHAPTER REVIEW ANSWERS

1. a
3. d
5. IES

Chapter **11**

MATERIALS, RENDERING, VISUAL STYLES, AND ANIMATION

In This Chapter

◇ Loading and assigning materials to solids and faces
◇ Material mapping
◇ Render and Render Region commands
◇ How to create your own visual style
◇ Creating an animation

11.1 INTRODUCING MATERIALS IN AUTOCAD

- Using materials in an AutoCAD scene involves several steps:
  - Select the desired material library from the Material Browser.
  - Drag the desired materials to be used.
  - Assign materials to objects or faces.
  - Assign a material mapping to different objects and faces.
  - Render to see the effects of your assignments.
- AutoCAD comes with a premade library of materials that contains (but is not limited to): Ceramic, Concrete, Flooring. To get the right results you should assign a material mapping to each face or object. A material mapping tells AutoCAD how to repeat the pattern and which angle to use.

11.2 MATERIAL BROWSER

- To start working with materials in AutoCAD, you have to start the Material Browser, which includes all the material definitions that come with AutoCAD. To issue this command, go to the **Render** tab, locate the **Materials** panel, and then select the **Material Browser** button:

- The Material Browser palette will be displayed as shown below:

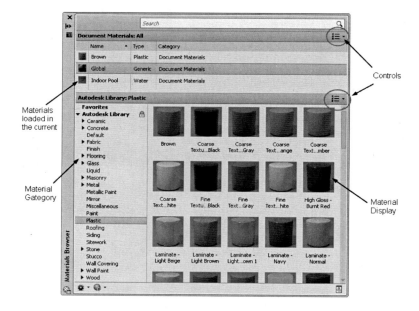

- As you can see, the Material Browser consists of three different parts:
    - The top part shows the loaded material in the current drawing.
    - The left part shows the existing material libraries (one of them will be the Autodesk Library as shown) and the categories holding the materials.
    - The right part shows the material display.

- The two controls at the left control how the different parts will be displayed. The following is the control of the top part of the Material Browser (it is self-explanatory):

- And this is the control for the materials at the bottom of the Material Browser:

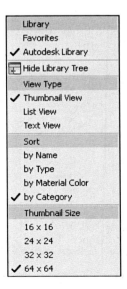

11.3 ASSIGNING MATERIALS TO OBJECTS AND FACES

- ▪ In general, there are three methods to assign materials to objects and faces:
  - Drag-and-drop
  - After selecting an object, right-click one material
  - Using the Attach by Layer command

11.3.1 Using Drag-and-Drop

- ▪ While the Material Browser is displayed, you should select the desired library, category, and material from the right and drag it and drop it on the desired object (solid, surface, or mesh). To assign the material to a single face, hold the [Ctrl] key when assigning, and only the selected face will be assigned the material.

11.3.2 After Selecting an Object, Right-Click the Material

- ▪ This method includes the selection of an object(s) (or face(s)) as the first step. The second step is to right-click on the desired material after you load it into the current file (using the drag-and drop method), and you will see the following menu. Select the **Assign to Selection** option, and the material will be assigned:

11.3.3 Using the Attach by Layer Command

- ▪ In order for this method to work, you should first make sure that all of the 3D objects are in the right layer. If this is the case, then issue the command by going to the **Render** tab, locating the **Materials** panel, and then selecting the **Attach By Layer** button:

- The following dialog box will be shown, which includes material loaded in the current drawing at the left and layers at the right:

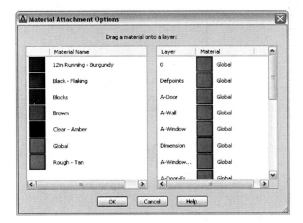

- Drag the material from the left side and then drop it on the layer on the right side. The following is a possible result:

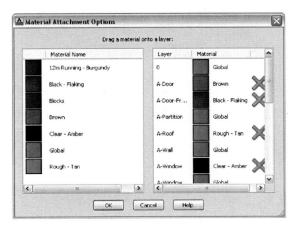

11.3.4 Removing Assigned Materials

- This command allows you to remove the assigned material from the object. In order to issue this command, go to the **Render** tab, locate the **Materials** panel, and then select the **Remove Materials** button:

- You will see the following prompt:

```
Select objects:
```

- And the mouse will change to the following shape:

- Select the desired object(s). When done press [Enter] to end the command.

11.4 MATERIAL MAPPING

- A material mapping answers the following questions:
  - What is the shape of your object (planar, cylindrical, or spherical)?
  - How many times will your image be repeated in columns and rows?
  - The value of the angle of the image.
  - The location of the image to start repeating.
- With a material mapping AutoCAD will produce rendered images that fit your needs exactly. To issue this command, go to the **Render** tab, locate the **Materials** panel, and then select the **Material Mapping** button:

- As shown, there are four types of material mapping:
  - Planar
  - Box
  - Cylindrical
  - Spherical

11.4.1 Planar Mapping

- This option will work on both planar faces and objects, but it is essentially for faces. You will see the following prompt:

```
Select faces or objects:
```

- Select the desired face, and you will see a green frame with four triangles at the four corners of the frame, something like the following:

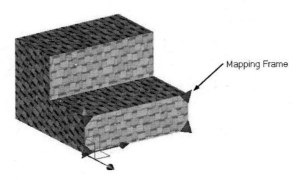

Mapping Frame

- Accordingly, the following prompt will be shown:

```
Accept the mapping or [Move/Rotate/reseT/sWitch mapping mode]:
```

- You can resize, move, and rotate the mapping. The following example shows how to resize the mapping:

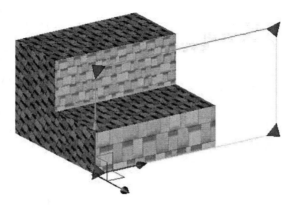

- Compare the resized pattern to the other parts of the object.
- See the following example with rotating:

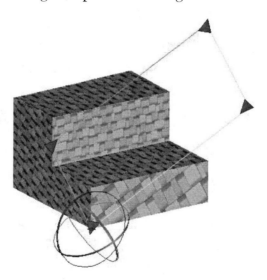

- Notice the pattern and compare it to the other parts of the object. When done press [Enter] to end the mapping process.

11.4.2 Box Mapping

- Designed to be used for box-like objects. The following prompts will be shown:

```
Select faces or objects:
```

- You will see a mapping box, like the following:

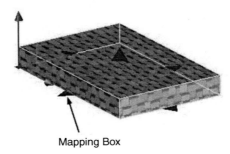

Mapping Box

- There are five triangles that control the size. Using these triangles allows you to resize, rotate, and move the box mapping. The following prompt will be shown:

```
Accept the mapping or [Move/Rotate/reseT/sWitch mapping mode]:
```

- This is the same prompt we dealt with for the planar mapping.
- See the following example:

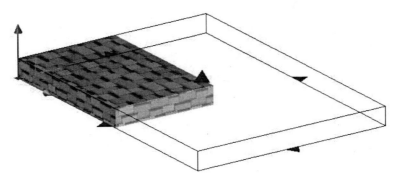

- Compare this illustration with the one before box mapping. When done press [Enter] to end the command.

11.4.3 Cylindrical and Spherical Mapping

- These are designed for cylindrical and spherical shapes. They will use the same prompts. The following help clarify the concept:

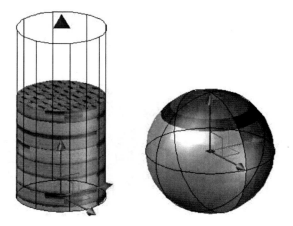

- As you can see, the cylindrical mapping has a single triangle that allows you to decrease or increase the mapping along the height, and the spherical mapping has no triangles, but has the ability to move the pattern towards the three main directions.

11.4.4 Copying and Resetting a Mapping

- To help you expedite the mapping process, AutoCAD allows you to do two things:
  - Copy an existing mapping from one object to another.
  - Reset the mapping to the default to start your mapping again.
- In order to issue these two commands, go to the **Render** tab, locate the **Materials** panel, and then select one of the two buttons as shown:

WORKING WITH MATERIALS

Practice 11-1

1. Start AutoCAD 2013.
2. Load the following materials from the Material Browser:

Category	Material
Wood	Andiroba
Metal	Brass-Polished Brushed-Blue-Silver
Glass	Clear-Amber Clear-Light
Stone	Coarse Polished-White
Sitework	Grass-Dark Rye
Flooring	Maple-Rosewood
Masonry	Modular-Light Brown
Fabric	Plaid 4 Stripes-Yellow-White

3. Assign materials to the layers using the following table:

Material	Layer
Andiroba	A-Door/A-Door-Frame/A-Window-Frame
Brass-Polished	Table_legs
Brushed-Blue-Silver	Lamp_Rod
Clear-Amber	Table_Top
Clear-Light	A-Window
Coarse Polished-White	Lamp_Base
Grass-Dark Rye	OuterBase
Maple-Rosewood	InnerBase
Modular-Light Brown	A-Wall
Plaid 4	Lamp_Cover
Stripes-Yellow-White	Couch

4. To see the effects of light and materials change the visual style to Realistic.
5. Turn the Sunlight on.
6. Take a look at the new look of your file.
7. Save and close.

11.5 INTRODUCING RENDER IN AUTOCAD

- To activate the camera, lights, and materials, you should use the **Render** command. The Render command will produce a graphical output of your model using the camera position, showing light effects along with shadows, and all the assigned materials. You have the ability to save the output to a graphical file along with controlling the graphical resolution of the file.
- There are two AutoCAD Render commands:
  - Render command
  - Render Region command

11.5.1 Render Command

- This command will render the entire scene shown on your screen. It will automatically open up a new window called **Render**. You will see the output produced gradually on the screen (the speed of the render depends on many factors, such as the complexity of the model, the amount of RAM available on your machine, the processor speed, etc.) using a **Render Preset,** which will be discussed shortly. You are allowed to zoom in and out using the wheel of the mouse. The **File** menu at the top of the screen can be used to save the output rendered image to a file (*.tif, *.png, *.jpg, etc.). To issue this command, go to the **Render** tab, locate the **Render** panel, and then select the **Render** button:

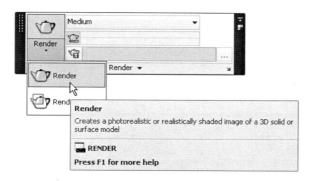

- You will see something like the following:

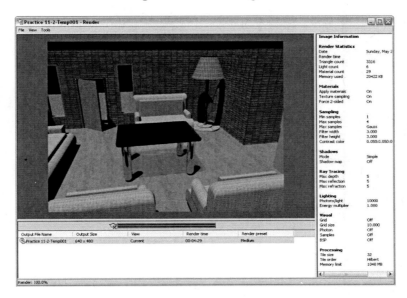

11.5.2 Render Region Command

- This command allows you to render the rectangular part of the scene. This will help you in two ways:
    - The time consumed is less than the time used in the Render command.
    - If you changed a light or a material and you want to test the effects, this command is your best choice.
- No window will pop up; everything will take place at your current display. To issue this command, go to the **Render** tab, locate the **Render** panel, and then select the **Render Region** button:

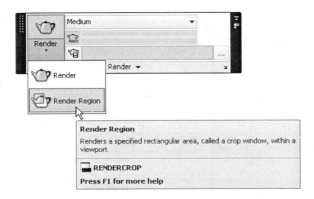

- You will see something like the following:

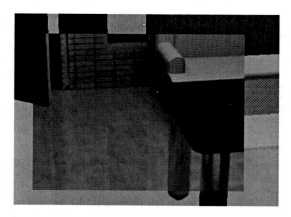

11.6 WHAT ARE RENDER PRESETS?

- Controlling all the parameters affecting the Render command has proven to be cumbersome. So, in order to make your life easier, AutoCAD created Render Presets, which include settings ready to use. They include:
  - Draft
  - Low
  - Medium
  - High
  - Presentation
- To select your desired Render Preset, go to the **Render** tab, locate the **Render** panel, and then select the **Render Preset**:

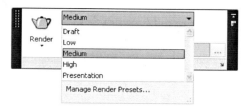

11.7 MANIPULATING THE OTHER RENDER SETTINGS

- While you are in the **Render** tab and the **Render** panel, you will find the following settings to edit:

- Render Quality
- Render Output Size
- Adjust Exposure
- Environment
- Render Window

11.7.1 Render Quality

- This option will help you adjust the rendered quality of the model. To change the render quality, go to the **Render** tab, locate the **Render** panel, and then select the **Render Quality** slider:

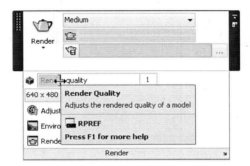

11.7.2 Render Output Size

- If your image is unclear, you may need to increase the render output size. But while you may get a better image, remember that the render time will also increase significantly and so will the file size. To change the image's resolution, go to the **Render** tab, locate the **Render** panel, and then select the Render Output Size pop-up list:

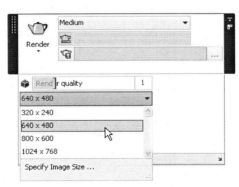

11.7.3 Adjust Exposure

- This command allows you to make adjustments to the rendered view, without reissuing the Render command. This command will allow you to change any or all of the following settings:
  - Brightness
  - Contrast
  - Mid tones
  - Allow Exterior Daylight?
  - Process Background?
- To issue this command, go to the **Render** tab, locate the **Render** panel, and then select the **Adjust Exposure** button:

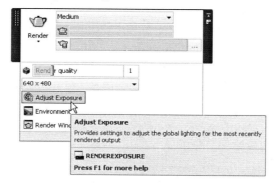

- The following dialog box will appear:

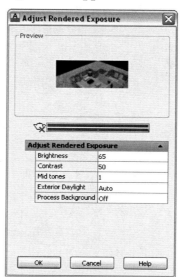

- Make the necessary changes and then click the **OK** button.

11.7.4 Environment

- This command allows you to include fog effects in the scene. To issue this command, go to the **Render** tab, locate the **Render** panel, and then select the **Environment** button:

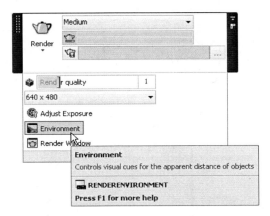

- The following dialog box will appear:

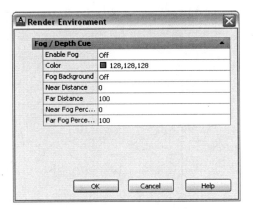

- You can control all or any of the following:
  - Turn fog on/off
  - Change the fog color
  - Turn fog background on/off
  - Near and far distances fog will cover
  - Near and far fog percentage

11.7.5 Render Window

- If you render and then close the window, you can see it again by using this command. To issue this command, go to the **Render** tab, locate the **Render** panel, and then select the **Render Window** button:

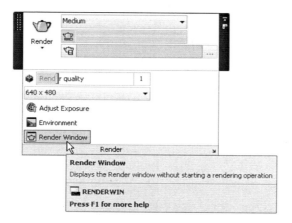

WORKING WITH RENDERING

Practice 11-2

1. Start AutoCAD 2013.
2. Open **Practice 11-2.dwg**.
3. Make sure you are using the Realistic visual style to see the lights and materials.
4. Select the camera nearest to you (Name = Camera 2), right-click, and select the Set Camera View option.
5. Make sure that Sun is on and Render Preset is Medium.
6. Using the Render Region command, select a region and render it (if you didn't install the Medium Image Library then select the option "Work without using the Medium Image Library") and see the shadows of the output.
7. Using the Render command, render the scene and name the file *Inside_using_Sun_Light.JPG*.
8. Turn off the Sunlight.
9. Change the Render Preset to Presentation.
10. Using the Render command, render the scene and name the file *Inside_night_Scene.TIF*.

11. If time permits and you have a well-equipped computer, you can render using a higher render quality and higher render output size. You can also use the other camera set in the file or input your own camera.
12. Save and close the file.

11.8 WHAT IS A VISUAL STYLE AND HOW DO YOU CREATE ONE?

- AutoCAD comes with ten premade visual styles:
  - 2D Wireframe: you will see all the lines whether they are hidden or not. This visual style will show the linetype and lineweight.
  - Conceptual
  - Hidden
  - Realistic
  - Shaded
  - Shaded with Edges
  - Shades of Gray
  - Sketchy
  - Wireframe: you will see all lines whether they are hidden or not. It will not show the linetype and lineweight.
  - X-Ray
- As discussed in Chapter 1, the easiest way to change the visual style is at the in-canvas viewport control at the upper-left corner of the screen. See the following:

- Though the existing ten premade visual styles are adequate for most purposes, there are times when you might want to create your own visual style. To create a new visual style, use the same list, but this time, select the last option, **Visual Styles Manager**:

- The Visual Style Manager palette will appear:

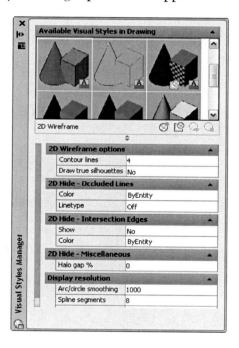

- Click the **Create New Visual Style** button to create a new visual style:

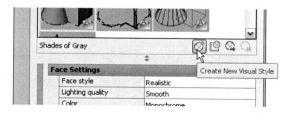

- The following dialog box will appear:

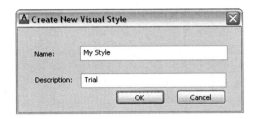

- Type the name of the new visual style and a proper description, then click the **OK** button. A new visual style will be added and ready for your manipulation. Change all or any of the following settings as discussed.

11.8.1 Face Settings

- The following are the face settings:

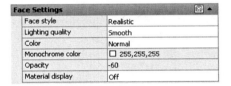

- Change the **Face Style** by selecting one of the following:
  - Realistic: realistic face style; the same as the Realistic visual style.
  - Gooch: warm-cool face style; the same as the Conceptual visual style.
  - None: same as the two Wireframes and Hidden visual styles.
- Change the **Lighting quality** by selecting one of the following:
  - Faceted: A facet is part of a face. Facet lighting will highlights facets.
  - Smooth: This is the default option, which will show smooth lighting on objects.
  - Smoothest: This is the best lighting setting, but depending on other settings, it may be not effective.
- Change the **Color** by selecting one of the following:
  - Normal: will display color of faces without any color modifier.
  - Monochrome: will display faces as black and white.
  - Tint: will display faces with monochrome color with some shading.
  - Desaturate: the same as tint but with a softer color.
- The value of the Monochrome color is grayed out. It will be enabled only when you change the Color setting to Monochrome.

- Change the value of **Opacity,** which specifies that the materials will be transparent or opaque. To turn Opacity on, click the following button:

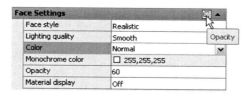

- Change **Material display** by selecting one of the following:
  - Materials and Textures
  - Materials
  - Off

11.8.2 Lighting

- The following are the lighting settings:

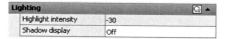

- Change the **Highlight intensity,** which controls the size of the highlight in faces without materials. In order to control this parameter, you have to unlock it first:

- Change **Shadow display,** which controls whether to show/hide the shadow in the visual style. The available choices are:
  - Mapped Object Shadows
  - Ground Shadows
  - Off

11.8.3 Environment Settings

- This option allows you to control whether to display a background or not:

11.8.4 Edge Settings

- The following are the edge settings:

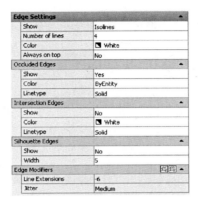

- Control **Show** and select one of the following options:
  - Facet Edges (control the color)
  - Isolines (control the number of lines, color, and whether to make the lines always on top)
  - None
- The following illustrates the concept:

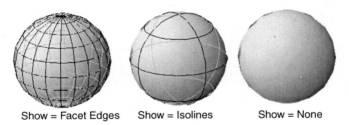

Show = Facet Edges   Show = Isolines   Show = None

- Control **Occluded Edges** (works only if Edge = Facet or Isolines), which controls whether to show/hide the hidden edges and their color and linetype. The following illustrates the concept:

Occluded Edges = No   Occluded Edges = Yes

- Control **Intersection Edges** (works only if Edge = Facet or Isolines), which specifies the color and linetype of the intersection line of multiple objects. The following illustrates the concept:

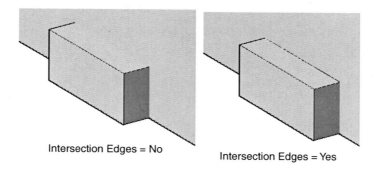

Intersection Edges = No   Intersection Edges = Yes

- Control **Silhouette Edges,** which specifies whether to show/hide them and specifies the width of the outer edges of an object. An acceptable silhouette value is 1–25. The following illustrates the concept:

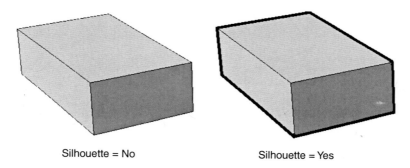

Silhouette = No   Silhouette = Yes

- Control **Edge Modifiers,** which mimic hand-drawn lines. In order to control these parameters you have to unlock them first. See the following:

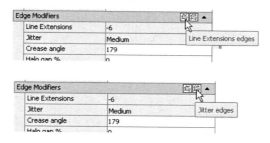

- Input values for extension and jitter (High, Medium, Low, and Off). The following illustrates the concept:

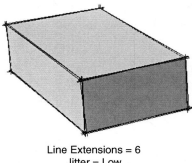

Line Extensions = 6
Jitter = Low

- After you finish editing the parameters you can ask AutoCAD to apply these settings to the current viewport:

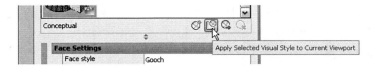

CREATING VISUAL STYLES

Practice 11-3

1. Start AutoCAD 2013.
2. Open **Practice 11-3.dwg**.
3. Create a new visual style and call it "Special."
4. Change the following parameters:
   a. Under Face Settings:
   - Face Style = Gooch
   - Lighting Quality = Faceted
   - Color = Monochrome
   b. Under Edge Settings:
   - Show = Facet Edges
   - Occluded = No
   - Intersection Edges = Yes
5. Apply these settings to the current viewport.

6. Make a comparison between the new style and Conceptual.
7. Save and close the file.

11.9 ANIMATION

- When you create an animation, you create a movie of a walkthrough or fly-by of a model. You can do this by using the Animation Motion Path command.
- In order to succeed using this method, you should first draw a camera path (using a line, polyline, spline, etc.) and a target path. You may use points instead, but with paths the output is more compelling.
- To issue this command, go to the **Render** tab, locate the **Animations** panel, and then select the **Animation Motion Path** button:

- The following dialog box will appear:

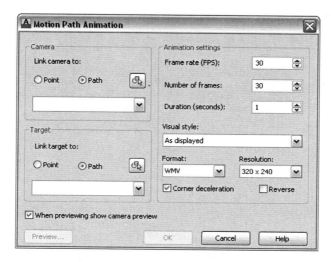

- Under **Camera**, you can attach the camera to a **Point** or a **Path**. Use the small button at the right to select either one. Under **Target** do the same. This step is optional. Under **Animation settings**, change all or any of the following:
  - Input **Frame rate (FPS—Frame Per Second)**. The default value is 30 FPS.
  - Input total **Number of frames**.
  - Input **Duration** in seconds.
- The above three values will affect each other. Knowing two will lead to the third automatically. Specify the visual style to be used in the animation. The list contains rendering options. If one of the rendering options is selected, it will take a long time to produce the movie.
- Specify the file **Format** of the output file. You have the ability to choose one of the four file formats:

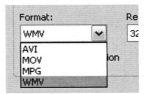

- Specify the **Resolution** of the output file (a higher resolution translates to a longer higher movie production time):

- Specify whether to reduce the speed of movement around corners (Corner deceleration). Finally, select to record the movie in reverse order starting from the end of the path.
- Click **Preview** to see a preview of the movie before it begins recording.
- To wrap it up, click **OK**, and AutoCAD will ask you to specify a name and location of the movie file, then it will start producing it. It may take few minutes or a few hours depending on your choices.

CREATING AN ANIMATION

Practice 11-4

1. Start AutoCAD 2013.
2. Thaw layer Target Path.
3. Start the Animation Motion Path and specify the camera point as follows:

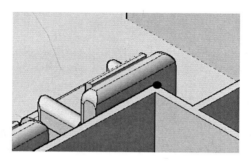

4. Set the target path to the polyline thawed.
5. Set the total to 10 seconds.
6. Set visual style to Realistic.
7. Set the file format to AVI.
8. Set the resolution to 640x480.
9. Name the file "Fixed Camera."
10. Freeze layer Target Path and thaw layer Camera Path.
11. Link the camera to the path and link the target to None.
12. Set the duration of the video to 20 seconds.
13. Set the file format to MPG, leaving all the other settings as is.
14. Name the file "Moving Camera."
15. Compare the AVI file size to the MPG file size. What do you think?
16. Save and close the file.

NOTES:

CHAPTER REVIEW

1. There are three methods to assign materials to objects, and one of them is "Attach By Layer."
    a. True
    b. False
2. Which one of the following is *not* related to visual style settings?
    a. Face color
    b. Intersection Edge color
    c. Highlight size
    d. Face style
3. Which one of the following is *not* related to Render in AutoCAD?
    a. There are two types of rendering in AutoCAD.
    b. You can control render output size.
    c. You can control render output file size.
    d. You can render using different presets.
4. MPG, AVI, and ASF are among the movie file formats AutoCAD supports.
    a. True
    b. False
5. _____ allows you to render a rectangular part of a scene.
6. _____ is the visual style that will mimic the human way of drawing.

CHAPTER REVIEW ANSWERS

1. a
3. c
5. Render Region

INDEX